T0121287

Naval Engineering

Principles and Theory of Gas Turbine Engines

Dennis L. Richardson

authorHOUSE®

AuthorHouse™
1663 Liberty Drive
Bloomington, IN 47403
www.authorhouse.com
Phone: 1 (800) 839-8640

Published by AuthorHouse 11/03/2016

ISBN: 978-1-5246-4857-2 (sc)
ISBN: 978-15246-4856-5 (e)

Library of Congress Control Number: 2016917752

Print information available on the last page.

http://www.navy.mil/management/photodb/photos/150511-N-FQ994-154.JPG
http://www.navy.mil/management/photodb/photos/101023-N-6632S-146.jpg
http://www.navy.mil/management/photodb/photos/070802-N-0000X-001.jpg
http://www.navy.mil/management/photodb/photos/161002-N-CS953-017.JPG

ACKNOWLEDGEMENT

This information contained herein has been adapted from the *Gas Turbine System Technician (Mechanical) 1 & C*, Volume 2, NAVEDTRA 10549, prepared by the Naval Education and Training Program Management Support Activity, 1987 and Gas Turbine System Technician (Mechanical) 3 &2, NAVEDTRA 10548, prepared by the Naval Education and Training Program Development Center, 1988; U.S. Government Printing Office Washington, D.C. 20402. To the extent, this book may contain text in the public domain; the Author makes no claim of ownership. The Author is credited with text compilation and editing. United States Navy cover photographs were taken by Tim Comerford, Robert Price, Jim Markle and Kevin J. Steinberg and released to the public.

CONTENTS

PREFACE

N*aval Engineering: Principles and Theory of Gas Turbine Engines* is organized to give you a systematic understanding and to serve as one of several sources of information for professional engineers and technical specialist. Operating, maintaining, and repairing the ship's propulsion plant and support system equipment is a job of vital importance. It requires teamwork and a special kind of supervisory ability. After reading this book, Engineers who have a high degree of technical competence and a deep sense of personal responsibility can develop this special ability.

Those who are assigned duties aboard a navy ship as engineers in the Navy Occupational Specialty (NOS) field of gas turbine specialist are expected to know and understand the information contained in this publication. The degree of success of the Navy depends (in part) on the engineer's ability and the manner in which they perform their assigned tasks in a wide variety of main propulsion, auxiliary, and electrical equipment.

Naval Engineers demonstrate technical leadership when orders are followed exactly, when safety precautions are followed, and when responsibility is accepted. Naval Engineers must continue to strive to improve leadership ability and technical knowledge through study, observation, and practical application. This book will surely assist.

CHAPTER 1

UNDERSTANDING GAS TURBINE ENGINES

This chapter is written with the intent to help you understand the history and development of Gas Turbine engines. We will review the thermodynamic processes known as the Brayton cycle of a gas turbine engine. Additionally, we will discuss various gas turbine nomenclatures, technical designs, applications, and performance conditions that affect the capabilities and limitations of marine operations. After reading this chapter, you will be proficient in describing the principal components of gas turbines and their construction.

History and Past Developments

Over the years, it has virtually been impossible to separate gas turbine technology and jet engine technology. Not until recently, when renowned professionals in both fields were able to apply sciences to both types of engines. However, the jet engine has been used more as a part of aviation.

The gas turbine has been used in many applications, including electric generation, ship propulsion, and even in the automobile industry with experimental propulsion. Today, many operational turbine power plants utilize a standardize aircraft jet engine as a Gas Generator (GG) with a Power Turbine (PT) and transmission added to complete the power plant.

In nature, the squid was using jet propulsion long before our science thought of it. There were examples of the reaction principle in early history; however, practical application of the reaction principle has occurred only recently. This delay is due to slow progress of technical achievement in engineering, fuels, and metallurgy (the science of metals).

A scientist by the name of Hero in Alexandria, Egypt described what is considered to be the first jet engine. Many sources have credited him as the inventor; true or not, the aeolipile is mentioned in multiple sources dating back as far as 250 B.C. There are many other examples of scientist throughout the course of history that used the principal of expanding gases to perform work; Leonardo da Vinci and Giovanni Branca are among this elite group.

Sir Isaac Newton described the laws of motion in the 1650s and later illustrated an example of the reaction principal in his famous steam wagon. And so, all devices that use the theory of jet propulsion are based on these laws. It was not long before John Barber, an Englishman, submitted the first patent for a design that used the thermodynamic cycle of the modern gas turbine (jet propulsion).

Modern Development

The patented application for the gas turbine was submitted in 1930 by another Englishman, Sir Frank Whittle. This particular patent was for application of a jet aircraft engine. Whittle used his own ideas along with the contributions of two other scientists such as Coley and Moss. After several failures, Whittle came up with a functional gas turbine engine (GTE).

American Development

The United States did not enter the gas turbine field until late in 1941. By then, General Electric was awarded a contract to build an American version of a foreign-designed aircraft engine. The engine and airframe were built in 12 months, then installed and subsequently used in the first jet aircraft flown in October 1942.

In late 1941, Westinghouse Corporation was awarded a contract to design and build the first all-American GTE. Their engineers designed the first axial flow compressor and annular combustion chamber. Both of these ideas, with minor changes, are the basis for the majority of contemporary engines in use today.

Marine Gas Turbines

The concept to utilize a gas turbine as the prime mover; propelling a United States Ship dates back to the late 1930s. At that time, a Pescara free piston gas engine was used experimentally with a gas turbine. The free piston engine (or gasifier) is a form of diesel engine. It uses air cushions instead of a crankshaft to return the pistons. It was an effective producer of pressurized gases. The German navy had previously used it in their submarines during World War II as an air compressor. In 1953, the French placed in service two small vessels powered by a free piston engine-gas turbine combination. In 1957 the Liberty ship William Patterson went into service on a trans-atlantic run. It had six free piston engines driving two turbines.

During that time, applications of the use of a rotary gasifier to drive a main propulsion turbine were used. The gasifier, or compressor, was usually an aircraft jet engine or turboprop front end. In 1947, the Motor Gun Boat 2009 of the British Navy used a 2500-hp gas turbine. It was used to drive the center of three shafts. In 1951 the tanker Auris, in an experimental application, replaced one of four diesel engines with a 1200-hp gas turbine. In 1956 the John Sergeant had a very efficient engine installed. It gave a fuel consumption rate of 0.523 pounds per hp/hr. The efficiency was largely due to use of a re-generator that recovered heat from the exhaust gases.

Later in the1950s, the marine gas turbine engine was becoming widely used, mostly by European Navies. All the applications combined the gas turbine plant with another conventional form of propulsion machinery. The gas turbine was used for high-speed operation while the conventional plant was used for cruising. The most common arrangements were the Combined Diesel or Gas (CODOG) or the Combined Diesel and Gas (CODAG) Systems. Diesel engines provided ships with good cruising range and reliability. But they have a disadvantage when used in anti-submarine warfare as their low-frequency sounds travel great distances through water. This made them easily detectable by passive sonar. Steam turbines have been combined to reduce low-frequency sound in the Combined Steam and Gas (COSAG) configuration like those used on the British Class Destroyers. However, the combination required more personnel to operate. Additionally, they did not have the long range of the diesel combinations. Another configuration that was very successful is the Combined Gas or Gas (COGOG) such as used on the British-type 42 DDG. These ships use the 4500-hp Tyne GTE for cruising and the Rolls Royce Olympus, a 28,000-hp engine for high speed.

The U.S. Navy entered the marine gas turbine field with the Asheville class patrol gunboats. These ships had the CODOG configuration with two diesel engines for cruising and the General Electric LM1500 gas turbine for high speed. The Navy has since designed and built destroyers, frigates, cruisers, and patrol hydrofoils that are entirely propelled by GTEs. This is a result of the reliability and efficiency of the advances in gas turbine designs.

Advantages and Disadvantages

The gas turbine, when compared to other types of engines, offers many advantages. Its greatest asset is its high power-to-weight ratio. This has made it, in the forms of turbo prop or turbo jet engine, the preferred engine for aircraft. Compared to the gasoline piston engine, the gas turbine operates on cheaper and safer fuel. The gasoline piston engine has the next best power-to-weight characteristics however the smoothness of the gas turbine, compared with reciprocating engines, has made it even more desirable in aircraft. Less vibration reduces strains on the airframe. In a warship, the lack of low-frequency vibration of gas turbines makes them preferable to diesel engines. There is less noise for a submarine to detect at long range. Modern production techniques have made gas turbines economical in terms of horsepower-per-dollar on initial installation. Their increasing reliability makes them a cost-effective alternative to steam turbine or diesel engine installation. In terms of fuel economy, modern marine gas turbines can compete with diesel engines. They may be superior to boiler/steam turbine plants when these are operating on distillate fuel.

However, there are definitely some disadvantages to gas turbines. Since they are high-performance engines, many parts are under high stress. Improper maintenance and lack of attention to details of maintenance procedure will impair engine performance. This may ultimately lead to engine failure. Something as simple as a pencil mark on a compressor turbine blade or a fingerprint can cause failure of the part. The turbine requires large quantities of air rendering the engine vulnerable to hazardous substances or foreign objects that can harm. Most gas turbine propulsion control systems are very complex because you have to control several factors. You have to monitor numerous operating conditions and parameters. The control systems must react quickly to turbine operating conditions to avoid casualties to the equipment. Gas turbines produce high-pitched loud noises which can damage the human ear. In shipboard installations special soundproofing is necessary. This adds to the complexity of the installation and makes access for maintenance more difficult. Also, the large amount of air used by a GTE requires large intake, exhaust ducting and filtration system. This takes up much valuable space on a small ship.

From a tactical standpoint, there are two major drawbacks to the GTE. The first is the large amount of exhaust heat produced by the engines. Most current anti-ship missiles are heat-seekers. The infrared (IR) signature of a gas turbine makes an easy target. Counter-measures are being developed to reduce this problem.

The second tactical disadvantage is the requirement for depot maintenance and repair of major casualties. The turbines are not overhauled in place on the ship. They must be removed and replaced by rebuilt engines if any major casualties occur. However, an engine can be changed wherever crane service and the replacement engine are available.

Future Trends

As Naval Engineers continue to improve materials and design applications, GTE will begin to be constructed to operate at higher combustion temperatures and pressures thus increasing the efficiency of gas turbine. Even now, gas turbine main propulsion plants offer fuel economy and installation costs no greater than diesel engines. Initial costs are lower than equivalent steam plants which typically burn distillate fuels. Future improvements have made gas turbines the best choice for non-nuclear propulsion of ships.

At present, marine gas turbines use aircraft jet engines for GGs. These are slightly modified for use in a marine environment, particularly in respect to corrosion resistance. As marine gas turbines continue to become more widely used, specific designs for ships may evolve. These compressors may be heavier and bulkier than aircraft engines and take advantage of re-generators to permit greater efficiency.

Probably large gas turbines cannot be made simple enough to overhaul completely in place. But progress is being made at doing major repairs in place. So they will require technical support from shore maintenance facilities. It is possible to airlift replacement engines so gas turbine ships can operate and be repaired worldwide on a par with their steam- or diesel-driven counterparts.

The high power-to-weight ratios of GTEs permit the development of high-performance craft such as hydrofoils and surface effect vehicles. These craft have high speed and are able to carry formidable weapons systems. They are being seen in increasing numbers in our fleet. In civilian versions, hydrofoils have been serving for many years to transport people on many of the world's waterways. Landing Craft, Air Cushion (LCAC) commonly referred to as Hovercraft are finding increased employment as carriers of people and support equipment. They are capable of speeds in excess of 30+ knots. If beach gradients are not too steep, they can reach points inland, marshy terrain, or almost any other level area.

Gas Turbine Operation

A gas turbine engine is composed of three major sections:

1. Compressor(s)
2. Combustion chamber(s)
3. Turbine wheel(s)

Here is a brief description of what takes place in a GTE during operation. Air is taken in through the air inlet duct by the compressor that compresses the air and thereby raises pressure and temperature. The air is then discharged into the combustion chamber(s) where fuel is admitted by the fuel nozzle(s). The fuel-air mixture is ignited by igniter(s) and combustion takes place. Combustion is continuous, and the igniters are de-energized after combustion is established. The hot and rapidly expanding gases are directed toward the turbine rotor assembly. Kinetic and thermal energy are extracted by the turbine wheel(s). The action of the gases against the turbine blades causes the turbine assembly to rotate. The turbine rotor is connected to the compressor, which rotates with the turbine. The exhaust gases then are discharged through the exhaust duct.

About 75 percent of the power development by a GTE is used to drive the compressor and accessories and 25 percent is used to drive a generator or to propel a ship through the water.

LAWS AND PRINCIPLES

To fully comprehend the basic engine theory, one must first be familiar with the physics concepts used in the operation of a GTE. In the following paragraphs we will cover the laws and principles that will apply to work. We will define, explain, and then demonstrate how they apply to a gas turbine.

Bernoulli's Principle. If an incompressible fluid flowing through a tube reaches a constriction, or narrowing of the tube, the velocity of fluid flowing through the constriction increases and the pressure decreases.

Boyle's Law. The volume of an enclosed gas varies inversely with the applied pressure, provided the temperature remains constant.

Charles' Law. If the pressure is constant, the volume of an enclosed dry gas varies directly with the absolute temperature.

Newton's Law. The first law states that a body at rest tends to remain at rest. A body in motion tends to remain in motion. The second law states that an imbalance of force on a body tends to produce an acceleration in the direction of the force. The acceleration, if any, is directly proportional to the force and inversely proportional to the mass of the body. Newton's third law states that for every action there is an equal and opposite reaction.

Pascal's Law. Pressure exerted at any point upon an enclosed liquid is transmitted undiminished in all directions.

Bernoulli's Principle

Consider the system illustrated in figure 1-1. Chamber A is under pressure and is connected by a tube to chamber B, which is also under pressure.

Chamber A is under static pressure of 100 psi. The pressure at some point, (X), along the connecting tube consists of a velocity pressure of 10 psi. This is exerted in a direction parallel to the line of flow. Added is the unused static pressure of 90 psi, which obeys Pascal's law and operates equally in all directions. As the fluid enters chamber B from the constricted space, it is slowed down. In so doing, its velocity head is changed back to pressure head. The force required to absorb the fluid's inertia equals the force required to start the fluid moving originally. Therefore, the static pressure in chamber B is again equal to that in chamber A. It was lower at intermediate point X.

The illustration (figure 1-1) disregards friction and is not encountered in actual practice. Force or head is also required to overcome friction. But, unlike inertia effect, this force cannot be recovered although the energy represented still exists somewhere as heat. Therefore, in an actual system the pressure in chamber B would be less than in chamber A. This is a result of the amount of pressure used in overcoming friction along the way.

At all points in a system the static pressure is always the original static pressure LESS any velocity head at the point in question. It is also LESS the friction head consumed in reaching that point. Both velocity head and friction represent energy that came from the original static head. Energy cannot be destroyed. So, the sum of the static head, velocity head, and friction at any point in the system must add up to the original static head. This, then, is Bernoulli's principle, more simply stated: If a non-compressible fluid flowing through a tube reaches a constriction, or narrowing of the tube, the velocity of fluid flowing through the constriction increases, and the pressure decreases. Bernoulli's principle governs the relationship of the static and dynamic factors concerning non-compressible fluids. Pascal's law governs the behavior of the static factors when taken by themselves.

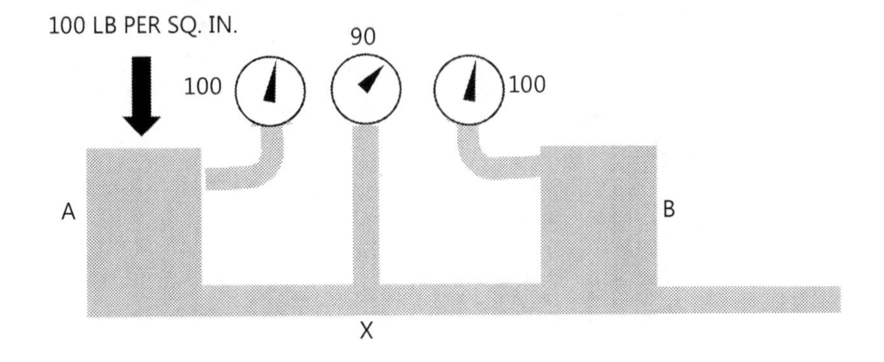

Figure 1-1.—Relation of static and dynamic factors—Bernoulli's principle.

Boyle's Law

Compressibility is an outstanding characteristic of gases. The English scientist, Robert Boyle, was among the first to study this characteristic. He called it the springiness of air. He discovered that when the temperature of an enclosed sample of gas was kept constant and the pressure doubled, the volume was reduced to half the former value. As the applied pressure was decreased, the resulting volume increased. From these observations he concluded that for a constant temperature the product of the volume and pressure of an enclosed gas remains constant. This became Boyle's law, which is normally stated: The volume of an enclosed dry gas varies conversely with its pressure, provided the temperature remains constant.

You can demonstrate Boyle's law by confining a quantity of gas in a cylinder which has a tightly fitted piston. Then apply force to the piston so as to compress the gas in the cylinder to some specific volume. If you double the force applied to the piston, the gas will compress to one half its original volume.

Changes in the pressure of a gas also affect the density. As the pressure increases, its volume decreases; however, there is no change in the weight of the gas. Therefore, the weight per unit volume (density) increases. So it follows that the density of a gas varies directly as the pressure, if the temperature is constant.

Charles' Law

Jacques Charles, a French scientist, provided much of the foundation for the modern kinetic theory of gases. He found that all gases expand and contract in direct proportion to the change in the absolute temperature. This is provided the pressure is held constant.

Any change in the temperature of a gas causes a corresponding change in volume. Therefore, if a given sample of a gas were heated while confined within a given volume, the pressure should increase. Actual experiments found that for each 1 °C increase in temperature, the increase in pressure was about 1/273 of the pressure at 0°C. Thus, it is normal practice to state this relationship in terms of absolute temperature. Basically, with a constant volume, the absolute pressure of an enclosed gas varies directly with the absolute temperature.

Examples of Charles' law: A cylinder of gas under a pressure of 1800 psig at 20 °C is left out in the sun. It heats up to a temperature of 55 °C. What is the new pressure within the cylinder? (You must convert the pressure and temperature to absolute pressure and temperature.)

P1 T1 =
P2 T2
or

1814.7 psia 293 °C absolute 2031.47 psia 328 °C absolute

Newton's First Law

Newton's first law states that a body at rest tends to remain at rest. A body in motion tends to remain in motion. This law can be demonstrated easily in everyday use. For example, a parked automobile will remain motionless until some force causes it to move—a body at rest tends to remain at rest. The second portion of the law—a body in motion tends to remain in motion—can be demonstrated only in a theoretical sense. The same car placed in motion would remain in motion (1) if all air resistance were removed, (2) if no friction were in the bearings, and (3) if the surface were perfectly level.

Newton's Second Law

Newton's second law states that an imbalance of force on a body tends to produce acceleration in the direction of the force. The acceleration, if any, is directly proportional to the force. It is inversely proportional to the mass of the body. This law can be explained by throwing a common softball. The force required to accelerate the ball to a rate of 50 ft/sec2 would have to be doubled to obtain an acceleration rate of 100 ft/sec2. However, if the mass of the ball were doubled, the original acceleration rate would be cut in half. You would have 50 ft/sec2 reduced to 25 ft/sec2. Do not confuse mass with weight. This law can be explained mathematically (A = acceleration; F = force; M = mass) $A = F/M$

Newton's Third Law

Newton's third law states that for every action there is an equal and opposite reaction. You have demonstrated this law if you have ever jumped from a boat up to a dock or a beach. The boat moved opposite to the direction you jumped. The recoil from firing a shotgun is another example of action-reaction. We can demonstrate this law with the same factors used in the second law in the equation ($F = MA$).

In an airplane, the greater the mass of air handled by the engine, the more it is accelerated by the engine. The force built up to thrust the plane forward is also greater. In a gas turbine, the thrust velocity can be absorbed by the turbine rotor and converted to mechanical energy.

This is done by adding more and progressively larger Power Turbine (PT) wheels.

Basic Engine Theory

Two factors are required for proper operation of a gas turbine. One is expressed by Newton's third law. The other is the convergent-divergent process. Convergent means approaching nearer together, as the inner walls of a tube that is constricted. Divergent means moving away from each other, as the inner walls of a tube that flares outward.

Bernoulli's principle is used in this process. The venturi of the common automobile carburetor is a common example of Bernoulli's principle and the convergent-divergent process. The following is a description of a practical demonstration of how a gas turbine operates. A blown-up balloon does nothing until the trapped air is released. The air escaping rearward causes the balloon to move forward (Newton's third law). If you could keep the balloon full of air, the balloon would continue to move forward. If a fan or pinwheel is placed in the air stream, the pressure energy and velocity energy will cause it to rotate. It can then be used to do work.

Now, replace the balloon with a tube or container (mounted in one place). Fill it with air from a fan or series of fans. They should be located in the air opening and driven by some source. You use the discharge air to turn a fan at the rear to do work. Imagine if fuel is added and combustion occurs, both the volume of air (Charles' law) and the velocity with which it passes over the fan are greatly increased. The horsepower the fan will produce is also increased. The continuous pressure created by the inlet fan, or compressor, prevents the hot gases from going forward.

Ironically if you attach a shaft to the compressor and extend it back to a turbine wheel, you have a simple gas turbine. It can supply power to run its own compressor and still provide enough power to do useful work. It could drive a generator or propel a ship.

Now, let's stop and recall the three basic parts of a gas turbine mentioned earlier:

1. Air is taken in through the air inlet duct by the compressor. There it is raised in pressure and discharged into the combustion chamber.
2. Fuel is admitted into the combustion chamber by the fuel nozzle(s). The fuel-air mixture is ignited by igniter(s) and combustion occurs.
3. The hot and rapidly expanding gases are directed aft through the turbine rotor assembly. There thermal and kinetic energy are converted into mechanical energy. The gases are then directed out through the exhaust duct.

Theoretical Cycle of GTE

Let's discuss a little on the cycle and theory of the gas turbine before we discussion on construction and design. As you are aware, a cycle is a process that begins with certain conditions. This cycle progresses through a series of additional conditions and returns to the original conditions.

The GTE operates on the ***Brayton Cycle***. The Brayton cycle is one where combustion occurs at constant pressure. Various components of the gas turbine were specifically designed to perform individual functions separately. These functions are known as Intake, Compression, Combustion, Expansion, and Exhaust.

The Brayton cycle can also be better explained as air entering the inlet at atmospheric pressure and constant volume. As the air passes through the compressor, it increases in pressure and decreases in volume. Combustion occurs at constant pressure while the increased temperature causes a sharp increase in volume. The gases at constant pressure and increased volume enter the turbine and expand through it. As the gases pass through the turbine rotor, the rotor turns kinetic and thermal energy into mechanical energy. The expanding size of the passages causes further increase in volume and a sharp decrease in pressure. The gases are released through the stack with a large drop in volume and at constant pressure. The cycle is continuous in a GTE, with each action occurring at all times.

Open and Closed Cycles

Most internal-combustion engines operate on an open engine cycle. This means the working fluid is taken in, used, and discarded. There are some gas turbines that operate on a semi-closed cycle. These engines use a re-generator much liked used on the gas turbine ship John Sergeant. However, the gas turbines you will encounter in the US Navy operate on the open cycle simply because open cycle offers the advantages of simplicity and light weight. Plus, all the working fluid only passes through the engine once.

The third classification of cycles is the closed cycle, where energy is added externally. The typical US Navy ship's steam plant is an example of a closed cycle system.

Convergent-Divergent Process

There are several pressure, volume, and velocity changes that occur within a gas turbine during operation. The convergent-divergent process is an application of Bernoulli's principle. (If a fluid flowing through a tube reaches a constriction or narrowing of the tube, the velocity of the fluid flowing through the constriction increases and the pressure decreases. The opposite is true when the fluid leaves the constriction; velocity decreases and pressure increases.) Boyle's law and Charles' law also come into play during this process. Boyle's law: The volume of any dry gas varies inversely with the applied pressure, provided the temperature remains constant. Charles' law:

If the pressure is constant, the volume of dry gas varies directly with the absolute temperature.

Now, let's review these laws as they apply to the gas turbine. Air is drawn into the front of the compressor. The rotor is specifically constructed that the area decreases toward the rear. This tapered construction gives a convergent area. Each succeeding stage is smaller, which increases pressure and decreases velocity (Bernoulli).

Between each rotating stage is a stationary stage or stator. The stator partially converts high velocity to pressure and directs the air to the next set of rotating blades. Because of its high rotational speed, the rotor imparts velocity to the air. Each pair of rotor and stator blades constitutes a pressure stage. Also, there is both a pressure increase at each stage and a reduction in volume (Boyle).

This process continues at each stage until the air charge enters the diffuser. There is a short area in the diffuser where no further changes take place. As the air charge approaches the end of the diffuser, you will notice that the opening flares (diverges) outward. At this point, the air loses velocity and increases in volume and pressure. Thus, the velocity energy has become pressure energy, while pressure through the diffuser has remained constant. The reverse of Bernoulli's principle and Boyle's law has taken place. The compressor continuously forcing more air through this section at a constant rate maintains constant pressure. Once the air is in the combustor, combustion takes place at constant pressure. After combustion there is a large increase in the volume of the air and combustion gases (Charles' law).

The combustion gases go rearward and occurs partially by velocity imparted by the compressor and partially because the rear area is a lower pressure. The end is the turbine nozzle section. Here you will find a decrease in pressure and an increase in velocity. The high-velocity, high-temperature, low-pressure gases are directed through the inlet nozzle to the first stage of the turbine rotor. The high-velocity, high-temperature gases cause the rotor to rotate by transferring velocity energy and thermal energy to the turbine blades. Turbine rotor is a divergent area. Between each rotating turbine stage is a static stage or nozzle. The nozzles act much the same as the stators in the compressor.

A nozzle is a stator ring with a series of vanes. They act as small nozzles to direct the combustion gases uniformly and at the proper angle to the turbine blades. Due to the design of the nozzles, each succeeding stage imparts velocity to the gases as they pass through the nozzle. Each nozzle converts heat and pressure energy into velocity energy by controlling the expansion of the gas. Each small nozzle has a convergent area.

Each stage of the turbine is larger than the preceding one. The pressure energy drops are quite rapid; consequently, each stage must be larger to use the energy of a lower pressure, lower temperature, and larger volume of gases. If more stages are used, the rate of divergence will be less.

The turbine motor area must diverge rapidly in proportion to the rate in which the air intake, compressor rotor and stator vanes (static) area converges into the diffuser area. Atmospheric air is raised in pressure and velocity and lowered in volume in area A by the compressor. Each stage can only compress air about 1.2 times, so the rate is limited. However, in the turbine rotor, the gases give up thermal and pressure energy and increase in volume through three stages. If this did not happen rapidly, back pressure from the turbine motor would cause the rear area to become choked. The gases in the combustor would back up into the compressor. There they would disrupt airflow and cause a condition known as surge, or compressor stall. This condition can destroy an engine in a matter of seconds. Surge will be explained later in our discussion of axial flow compressors (stators).

The gases from the last turbine stage enter the exhaust duct where they are transmitted to the atmosphere. The leading portion of the exhaust duct is part of a divergent area. Further divergence reduces the pressure and increases the volume of the warm gases and aids in lowering the velocity. The exhaust gases enter the atmosphere at or slightly above atmospheric pressure. This depends on the length and size of the exhaust duct.

Air enters the intake at constant pressure. It is compressed as it passes through the compressor.

Between the end of the diffuser and the beginning of the turbine nozzle section, combustion occurs and volume increases. As the gases pass through the turbine rotor area, the gases expand with a drop in pressure and an increase in volume. The gases are discharged to the atmosphere through the exhaust duct at constant pressure. This is how a simple gas turbine works.

Adiabatic Compression

In the ideal gas turbine, the air enters the compressor and is compressed adiabatically. In an adiabatic stage change there is no transfer of heat to or from the system during the process. In many real processes, adiabatic changes can occur when the process is performed rapidly. Since heat transfer is relatively slow, any rapidly performed process can approach an adiabatic state. Compression and expansion of working fluids are often achieved almost adiabatically.

During operation the work produced by the compressor turbine rotor is almost the same amount as the work required by the compressor. The mass flow available to the compressor turbine is about the same as the mass flow handled by the compressor. Therefore, the heat of compression will closely equal the heat of expansion. Allowances are made for factors such as bleed air, pressure of fuel added, and heat loss to turbine parts.

As the high-temperature, high-pressure gases enter the turbine section, they expand rapidly. There is relatively little change in temperature of the gases. The net power available from the turbine is the difference between the turbine- developed power and the power required to operate the compressor.

Ambient Temperature Effects

The power and efficiency of a GTE are affected by both outside and inside variables. Air has volume that is directly affected by its temperature. As the temperature decreases, the volume of air for a given mass decreases and its density increases. Consequently, the mass weight of the air increases which, in turn, increases efficiency. This happens because less energy is needed to achieve the same compression at the combustion chambers. Also, cooler air causes lower burning temperatures. The resulting temperatures extend turbine life. For example, a propulsion gas turbine is operating at 100 percent GG speed with 100 percent PT speed. The ambient (external air) temperature is 70 °F. If the temperature were increased to 120 °F, the volume of air would increase. The mass weight would decrease. Since the amount of fuel added is limited by the inlet temperature the turbine will withstand, the mass weight flow cannot be achieved; the result is a loss of net power available for work. The plant may be able to produce only 90 to 95 percent of its rated horsepower.

On the other hand, if the ambient temperature were to drop to 0°F, the volume of air would decrease. The mass weight would increase. Since the mass weight is increased and heat transfer is better at higher pressure, less fuel is needed to increase volume; the result is a heavier mass of air at the required volume. This situation produces quite an efficient power plant. It has a GG speed of 85 to 90 percent and a PT speed of 100 percent. In the case of a constant speed engine, the differences in temperature will show up on exhaust gas temperature. In some cases, it will show up on the load the engine will pull. For instance, on a hot day of 120 °F, the engine on a 300-kw generator set may be able to pull only 275 kw. This is due to limitations on exhaust or turbine inlet temperature. On a day with 0°F ambient temperature, the same engine will pull 300-kw. It can have an exhaust or turbine inlet temperature that is more than 100 °F, lower than average. Here again, less fuel is needed to increase volume and a greater mass weight flow. In turn, the plant is more efficient.

Compressor Cleanliness

Another factor that will have a great effect on performance is the condition of the compressor. A clean compressor is essential to efficiency and reliability. During operation at sea the compressor will ingest salt spray. Over a period of time, this salt will build up in the compressor. Salt buildup is relatively slow. It will occur more on the stator vanes and the compressor case than on rotating parts. Centrifugal force tends to sling salt contaminants off the rotor blades. Also, oil rapidly increases contamination of the compressor. Any oil ingested into the engine coats the compressor with a film. The film traps any dust and other foreign matter suspended in the air. The dust and dirt absorb more oil, which traps more dirt, and so forth. If left unattended, the buildup of contamination will lead to a choking of the compressor and a restricted airflow. In turn, gradually more fuel is required. So the gas temperatures will rise until loss of power and damage to the turbine may result. Contamination, if not controlled, can lead to a compressor surge during engine start. It will also reduce the life of the compressor and turbine blading.

Gas Turbine Engine Categories

There are several different types of gas turbines, all using the same basic principles already discussed. GTEs are classified by their construction (the type of compressor, combustor, or its shafting). The compressor may be either centrifugal or axial type. The combustor may be annular, can-annular, or can type. The type of shaft used on a GTE is also used to classify an engine. The three types are single shaft, split shaft, or twin spool. These classifications will be discussed on the following pages.

Compressor Classification

Gas turbines may be classified by compressor type, according to the direction of the flow of air through the compressor. The two principal types are centrifugal flow and axial flow. The centrifugal compressor draws in air at the center or eye of the impeller and accelerates it around and outward. In the axial flow engine the air is compressed while continuing its original direction of flow. The flow of air is parallel to the axis of the compressor rotor.

Centrifugal Compressor. The centrifugal compressor is usually located between the accessory section and the combustion section. The basic compressor section consists of an impeller, diffuser, and compressor manifold. The diffuser is bolted to the manifold. Often the entire assembly is referred to as the diffuser. For simplicity, we will refer to each unit separately.

The impeller may be either single entry or dual entry. The main differences between the single entry and dual entry are the size of the impeller and the ducting arrangement. The single entry impeller permits convenient ducting directly to the inducer vanes. The dual entry uses a more complicated ducting to reach the rear side. Single entry impellers are slightly more efficient in receiving air. They must be of greater diameter to provide sufficient air which increases the overall diameter of the engine.

Dual entry impellers are smaller in diameter. They rotate at higher speeds to ensure sufficient airflow. Most gas turbines of present-day design use the dual entry compressor to reduce engine diameter. The air must enter the engine at almost right angles to the engine axis. Because of this a plenum chamber is also required for dual entry compressors. The air must surround the compressor at positive pressure before entering the compressor to give positive flow.

Operating Principles

The compressor draws in the entering air at the hub of the impeller and accelerates it radially outward by means of centrifugal force through the impeller. It leaves the impeller at a high velocity low pressure and flows through the diffuser. The diffuser converts the high-velocity, low-pressure air to low velocity with high pressure. The compressor manifold diverts the flow of air from the diffuser, which is an integral part of the manifold, into the combustion chambers.

Construction. In the centrifugal compressor the manifold has one outlet port for each combustion chamber. The outlet ports are bolted to an outlet elbow on the manifold. The outlet ports ensure that the same amount of air is delivered to each combustion chamber.

The outlets are known by a variety of names. Regardless of the names used, the elbows change the airflow from radial flow to axial flow. Then the diffusion process is completed after the turn. Each elbow contains from two to four turning vanes to efficiently perform the turning process. They also reduce air pressure losses by presenting a smooth turning surface. The impeller is usually fabricated from forged aluminum alloy, heat-treated, machined, and smoothed for minimum flow restriction and turbulence. Some types of impellers are made from a single forging. In other types the inducer vanes are separate pieces. Centrifugal compressors may achieve efficiencies of 80 to 84 percent at pressure ratios of 2.5:1 to 4:1 and efficiencies of 76 to 81 percent at pressure ratios of 4:1 to 10:1.

Advantages: rugged, simple in design, relatively light in weight, and develops high-pressure ratio per stage.

Disadvantages: large frontal area, lower efficiency, and difficulty in using two or more stages due to air loss that will occur between stages and seals.

Axial Flow Compressors. There are two main types of axial compressors. One is the drum type and the other is the disk type. The purpose of the axial compressor is the same as the centrifugal type. Both take in ambient air and increase the velocity and pressure. They discharge the air through the diffuser into the combustion chamber. The two main elements of an axial flow compressor are the rotor and stator.

The rotor has fixed blades which force the air rearward much like an aircraft propeller. Behind each rotor stage is a stator. The stator directs the air rearward to the next rotor stage. Each consecutive pair of rotor and stator blades constitutes a pressure stage. The action of the rotor at each stage increases compression of the air at each stage and accelerates it rearward. By virtue of this increased velocity, energy is transferred from the compressor to the air in the form of velocity energy. The stators at each stage act as diffusers, partially converting high velocity to pressure. The number of stages required is determined by the amount of air and total pressure rise required. The greater the number of stages, the higher the compression ratio. Most present-day engines have 8 to 16 stages, depending on air requirements.

Compressor Construction. The rotor and stators are enclosed in the compressor case. Present-day engines use a case that is horizontally divided into upper and lower halves. The halves are normally bolted together with either dowel pins or fitted bolts. They are located at various points. They ensure proper alignment to each other and in relation to other engine assemblies. The other assemblies bolt to either end of the compressor case. On some older design engines, the case is a one-piece cylinder open on both ends. The one-piece compressor case is simpler to manufacture; however, any repair or detailed inspection of the compressor rotor is impossible. The engine must be removed and taken to a shop. There it is disassembled for repair or inspection of the rotor or stators. On many engines with the split case, either the upper or lower case can be removed. The engine can remain in place for maintenance and inspection.

The compressor case is usually made of aluminum or steel. The material used will depend on the engine manufacturer and the accessories attached to the case. The compressor case may have external connections made as part of the case. These connections are normally used to bleed air during starting and acceleration or at low-speed operation.

Drum-Type Construction. The drum-type rotor consists of rings that are flanged to fit one against the other. The entire assembly may then be held together by through bolts. The drum is one diameter over its full length. The blades and stators vary in length from front to rear. The compressor case tapers accordingly. This type of construction is satisfactory for low-speed compressors where centrifugal stresses are low.

Disk-Type Construction. The disk-type rotor consists of a series of disks constructed of titanium alloys, low alloy steel, and stainless steel. The blades vary in length from entry to discharge. This is due to a progressive reduction in the annular working space (drum to casing) toward the rear. The working space decreases because the rotor disk diameter increases. The disk-type rotors are used almost exclusively in all present-day, high-speed engines.

Compressor Blading. Each stage of an axial compressor consists of a set of rotor and stator blades. Stator blades may also be referred to as vanes. The construction of these blades is important to efficient operation of a gas turbine.

Rotor Blades. The rotor blades are usually machined from stainless steel forgings although some blades can be made of titanium. Methods of attaching the blades in the rotor disk rims vary in different designs. They are commonly fitted into disks by either bulb or fir-tree type of roots. The blades are then locked by means of grub-screws, peening, lock wires, pins, or keys.

The clearance between the rotating blades and the outer case is important with many manufacturers depending on a wear fit between the blade and the compressor case. Many companies design their blades with knife edge tips so that the blades will wear away and form their own clearance as they expand from the heat generated from compression of the air.

Another method of preventing excessive rubbing while maintaining minimum clearance is to metal-spray the case and stators. Thin squealer tips on the blades and vanes contact the sprayed material. The abrasive action of the blade tip prevents excessive rubbing while obtaining minimum clearance. The primary causes of rubbing are an excessively loose blade or a malfunction of a compressor support bearing. This causes the compressor rotor to drop.

Large compressors have loose-fitting blades on the first several stages. These move during blade called a midspan platform. The platform gives some radial support to the blades during acceleration. Support is needed because of the length and amount of movement of the blades and to prevent excessive blade vibration.

Stators. The stator vanes project radially toward the rotor axis. They fit closely on either side of each stage of the rotor. The function of the stators is twofold: (1) they receive air from the air acceleration to minimize vibration while passing through critical speed ranges. Once up to speed, centrifugal force locks the blades in place and little or no movement occurs. There is also movement of the blades during rundown. On a clean engine some of the blades may have as much as 1/4-inch radial movement. You may hear a tinkling sound during rundown.

Large compressor rotors have long blades on the first stage. They have a piece made onto the inlet duct or from each preceding stage of the rotor. They then deliver the air to the next stage or to combustors at a workable velocity and pressure; (2) the stators also control the direction of air to each rotor stage to obtain the maximum possible compressor blade efficiency. The stator vanes are usually made of steel with corrosion- and erosion-resistant qualities. Most of the vanes are shrouded by a band of suitable material to simplify the fastening problem and then welded into the shrouds. The outer shrouds are secured to the inner wall of the compressor case by radial retaining screws.

Some manufacturers machine a slot in the outer shrouds and run a long, thin key the length of the compressor case. The key is held in place by retaining screws that prevents the stators from turning within the case. This method is used when a one-piece compressor case is slid over the compressor and stator assembly.

Each pair of vanes in a stator acts as a diffuser. The vanes use the divergent principle; the outlet of the vane area is larger than the inlet. This diverging area takes the high-velocity, low-pressure air from the preceding rotor stage. It converts it into a low-velocity, high-pressure airflow. Then it directs it at the proper angle to the next rotor stage. The next rotor stage will restore the air velocity that was lost because of the pressure rise. The next stator will give a further pressure rise. This process continues for each stage in the compressor.

A pressure rise of about 1.2 times the preceding stage is as much as a single stage can handle. The higher pressure rises result in higher diffusion rates with excessive turning angles. This causes excessive air instability; low efficiency.

Preceding the stators and the first stage of the compressor is a row of vanes known as Inlet Guide Vanes (IGV). The function of the IGV varies somewhat, depending on the size of the engine and the air-inlet construction. On smaller engines the air inlet is not totally in line with the first stage of the rotor. The IGV straighten the airflow and direct it to the first-stage rotor. On large engines the IGV are variable and move with the variable stators. The variable IGV on large engines direct the airflow at the proper angle to reduce drag on the first-stage rotor. Variable IGV achieve the same purposes as Variable Stator Vanes (VSV).

The variable stators are controlled by Compressor Inlet Temperature (CIT) and engine power requirements. There moved by mechanical linkages that are connected to the fuel-control governor. Variable stators have a two-fold purpose: (1) they are positioned at various angles depending on compressor speed. They ensure the proper angle of attack between the compressor blades. Varying the blade angle helps to maintain maximum compressor efficiency over the operating speed range of the engine. This is important in variable speed engines such as those used for main propulsion; (2) the variable stators on large engines virtually eliminate compressor surge. Surge results when the airflow stalls across the compressor blades; that is, air is not smoothly compressed into the combustion and turbine section. Stalling may occur over a few blades or a section of some stages. If enough flow is interrupted, pressure may surge back through the compressor. This occurrence can be minor or very severe with resulting damage to the turbine.

All the air in the combustor then can be used for combustion instead of only the primary air. Lack of cooling air may cause extreme temperatures that could burn the combustor and turbine section. In short, by changing the angle of the stators and use of bleed valves, the airflow through the compressor is ensured. Compressor surge can be almost totally prevented. Constant-speed engines, such as those used to drive generators, normally do not use variable stators. They are designed to operate at 100-percent rpm all the time. The proper fuel schedule and bleed air valves are adequate to prevent or minimize compressor surge.

DEFINING GAS TURBINES BASED OFF COMBUSTION CHAMBER DESIGN

There are three types of combustion chambers: (1) Can type, (2) Annular type, and (3) Can-Annular type. The can-type chamber is used primarily on engines that have a centrifugal compressor. The annular and can-annular types are used on axial flow compressors.

Can-Type Chamber

The can-type combustion system consists of individual liners and cases mounted around the axis of the engine. Each chamber contains a fuel nozzle. This arrangement makes removing a chamber easy; however, it is a bulky arrangement and makes for a structurally weak engine. The outer casing is welded to a ring that directs the gases into the turbine nozzle. Each of the casings is linked to the others with a short tube. This arrangement ensures that combustion occurs in all the burners during engine start. Inside each of these tubes is a flame tube that joins an adjacent inner liner.

Annular-Type Chamber

The annular-type combustion chamber is usually found on axial flow engines. It is probably one of the most popular combustion systems in use. The construction consists of a housing and liner the same as the can type. The great difference is in the liner. On large engines, the liner consists of an undivided circular shroud extending all the way around the outside of the turbine shaft housing. A large one-piece combustor case covers the liner and is attached at the turbine section and diffuser section.

The dome of the liner has small slots and holes to admit primary air. They also impart a swirling motion for better atomization of fuel. There are also holes in the dome for the fuel nozzles to extend through into the combustion area. In the case of the double-annular chamber, two rows of fuel nozzles are required. The inner and outer liners form the combustion space. The outer liner keeps flame from contacting the combustor case. The inner liner prevents flame from contacting the turbine shaft housing. Additionally, there are large holes and slots strategically located along the liners to; (1) admit some cooling air into the combustion space to help cool the hot gases to a safe level, (2) center the flame, and (3) admit the balance of air for combustion. These gases are cooled enough to prevent warpage of the liners in the process.

The annular-type combustion chamber is a very efficient system that minimizes bulk. It can be used most effectively in limited space. There are some disadvantages, however. On some engines, the liners are one piece and cannot be removed without engine disassembly. Also, engines that use a one-piece combustor dome must be disassembled to remove the dome.

Can-Annular Type of Chamber

The can-annular type of combustion chamber combines some of the features of both the can and the annular burners. The design is a result of the split spool compressor concept due to problems that was previously encountered with a long shaft and with one shaft within the other. As a result, a can-annular chamber was designed to perform all the necessary functions.

In the can-annular type of chamber individual cans are placed inside an annular case. The cans are essentially individual combustion chambers with concentric rings of perforated holes to admit air for cooling. On some models each can has a round perforated tube which runs down the middle of the can. The tube carries additional air which enters the can through the perforations to provide more air for combustion and cooling. The effect is to permit more burning per inch of can length than could otherwise be done.

Now, fuel nozzle arrangement varies from one nozzle in each can to several nozzles around the perimeter of each can. The cans are inherent resistance to buckling because of their small diameter. Each can has two holes that are opposite each other near the forward end of the can. One hole has a collar called a flame tube. So when the cans are assembled in the annular case, these holes and their collars form open tubes. The tubes are between adjacent cans so that a flame passes from one can to the next during engine starting.

The short length of the can-annular type of chamber is a structural advantage. It provides minimal pressure drop of the gases between the compressor outlet and the flame area. Another advantage of the can-annular engine is the greater structural strength it gets from its short combustor area. Maintenance is also simple. Maintenance personnel can just slide the case back and remove any one burner for inspection and/or repair. Another good feature is the relatively cool air in the annular outer can. It tends to reduce the high temperatures of the inner cans while the air blanket keeps the outer shell of the combustion section cooler.

Classification of Gas Turbines by Type of Shafting

Several types of gas turbine shafts are used. These are Single Shaft, Split Shaft, and Twin Spool. Of these, the single shaft and split shaft are the most common in use in Naval vessels.

The single shaft engine concept is predominately used in the U.S. Navy applications to drive ship's service generators. The split shaft engine is used for main propulsion as a variety of speed ranges is encountered.

In the split shaft engine, there is no mechanical connection between the GG turbine and the PT. With this type of engine, the output speed can be varied by varying the generator speed. Also, under certain conditions, the GG can run at a reduced rpm and still provide maximum PT rpm. The reduced rpm greatly improves fuel economy and also extends the life of the GG turbine. The starting torque required is lowered simply because the PT, reduction gears, and output shaft are stationary until the GG reaches idle speed. An additional feature in the multi-shaft marine propulsion plant is that the GG rotates only one way. One design (clockwise rotation or counterclockwise rotation) of the GG can be used on either shaft; however, the PT can be made to rotate either way. This is achieved by changing the PT wheel and nozzles. The arrangement was used for propulsion gas turbines aboard the late DD-963 and FFG-7 class ships.

Another type of turbine is the twin spool, commonly referred to as a multi-stage gas turbine in where there are two separate compressors and two separate turbine rotors. They are referred to as Low-Pressure (LP) compressor and turbine rotor and High-Pressure (HP) compressor and turbine rotor. The LP compressor and turbine are connected by a shaft. This shaft runs through the hollow shaft that connects the HP turbine to the HP compressor. The starter drives the HP assembly during engine start. The PT functions the same as in the split shaft engine. A larger volume of air can be handled as compared to a single or split shaft engine; however, the engine has more moving parts. The increase in overall dimensions and complexity makes the engine less desirable for ship's propulsion than the split shaft engine.

TURBINE ASSEMBLIES

Gas turbine engines are not normally classified by turbine type. In theory, design, and operating characteristics, the turbines used in GTEs are similar to those used in a steam plant. The gas turbine differs from the steam turbine mainly in the type of blading material used, the means provided for cooling the turbine shaft bearings, and the lower ratio of blade length to wheel diameter.

The terms GG turbine and PT are used to differentiate between the turbines. The GG turbine powers the GG and accessories. The PT powers the ship's propeller through the reduction gear and shafting. The turbine that drives the compressor of a GTE is located directly behind the combustion chamber outlet. This turbine consists of two basic elements, the stator or nozzle, and the rotor.

Turbine Stators

The stator element of the turbine section is known by a variety of names but the three most common are turbine nozzle vanes, turbine guide vanes, and nozzle diaphragm. However, throughout our discussion turbine stators are usually referred to as nozzles. The turbine nozzle vanes are located directly aft of the combustion chambers and immediately forward of, and between the turbine wheels. Turbine nozzles have a two-fold function. First, the nozzles prepare the mass flow for harnessing of power through the turbine rotor. This occurs after the combustion chamber has introduced the heat energy into the mass airflow and delivered it evenly to the nozzles. The stationary vanes of the turbine nozzles are contoured and set at a certain angle. They form a number of small nozzles that discharge the gas as extremely high-speed jets; thus, the nozzle converts a varying portion of the heat and pressure energy to velocity energy. The velocity energy can then be converted to mechanical energy through the rotor blades.

The turbine nozzle functions to deflect the gases to a specific angle in the direction of turbine wheel rotation. The gas flow from the nozzle must enter the turbine blade passageway while it is still rotating. Therefore, it is essential to aim the gas in the general direction of turbine rotation. The turbine nozzle assembly consists of an inner shroud and an outer shroud between which are fixed the nozzle vanes.

The number of vanes varies with different types and sizes of engines; typically turbine nozzles featuring loose and welded vane fits.

The vanes of the turbine nozzle are assembled between the outer and inner shrouds or rings in different ways. Although turbine nozzles may differ in their construction, there is one characteristic special to all turbine nozzles; and that is, the nozzle vanes must be constructed to allow for thermal expansion. Otherwise, rapid temperature variances could cause distortion or warping of the metal components. Thermal expansion of turbine nozzles is accomplished by one of several methods. In one method, the vanes are assembled loosely in the supporting inner and outer shrouds in which each of the vanes fits into a contoured slot in the shrouds. The slots conform to the airfoil shape of the vanes. These slots are slightly larger than the vane to give a loose fit. For further support the inner and outer shrouds are encased by an inner and an outer support ring. This adds strength and rigidity to the turbine nozzle. These supports also permit removal of the nozzle vanes as a unit; otherwise, the vanes could fall out of the shrouds as the shrouds are removed. Another method to allow for thermal expansion is to fit the vanes into inner and outer shrouds. In this method the vanes are welded or riveted into position. Either the inner or the outer shroud ring is cut into segments to provide for thermal expansion. The saw cuts dividing the segments will allow enough expansion to prevent stress and warping of the vanes.

The basic types of construction of nozzles are the same for all types of turbines. The convergent-divergent principle (Bernoulli's principle) is used to increase gas velocity. The turbine nozzles are made of high-strength steel. Steel that can withstand the direct impact of the hot, high pressure, high-velocity gases from the combustor. The nozzle vanes must also resist erosion from the high-velocity gases passing over them. Increasing the inlet gas temperature by about 750 °F achieves almost a 100-percent increase in specific horsepower. However, nozzles do not stand up for long to the higher temperatures, hence the different methods of increasing nozzle attempted over the years. One method that was tried was to coat the nozzle with a ceramic coating. Higher temperatures were achieved. However, the different expansion rates of the steel and the ceramic caused the coating to break away over a period of time. Experiments are still being conducted, even so far as to use an entirely ceramic nozzle. Another means of withstanding high temperatures is to use newly developed alloys. However, extreme costs of the alloys prevented commercial production of such nozzles. The most widely use method today in large engines, is to use air-cooled nozzle vanes. Compressor bleed air is fed through passages to the turbine where it is directed to the nozzle. The air cools both the turbine and the nozzle. The nozzle may also be cooled by air admitted from the outer perimeter of the nozzle ring.

The nozzle vanes are made with many small holes or slots on the leading and trailing edges. Air is forced into the nozzle and out through the slots and holes. The vane is cooled as the air passes through. The air is discharged into the hot gas stream, passing through the remainder of the turbine section and out through the exhaust duct. Cooling air is used primarily in the HP turbine section. The temperature of the gases is at an acceptable level by the time the gases reach the LP turbine section. Seals installed between the nozzle entrance shroud and the turbine shaft is often pressurized with bleed air. This helps to minimize interstage leakage of the gases as they pass through the turbine.

Turbine Rotors

The rotor element of the turbine consists of a shaft and bladed wheel(s). The wheel(s) are attached to the main power transmitting shaft of the GTE. The jets of combustion gas leaving the vanes of the stator element act upon the turbine blades. Thus, the turbine wheel can rotate in a speed range of about 3600 to 42,000 rpm. The high rotational speed imposes severe centrifugal loads on the turbine wheel. At the same time the high temperature (1050° to 2300°F) results in a lowering of the strength of the material. Consequently, the engine speed and temperature must be controlled to keep turbine operation within safe limits. The operating life of the turbine blading usually determines the life of the GTE.

The turbine wheel is a dynamically balanced unit consisting of blades attached to a rotating disk. The disk in turn is attached to the rotor shaft of the engine. The high-velocity exhaust gases leaving the turbine nozzle vanes act on the blades of the turbine wheel. This causes the assembly to rotate at a very high rate of speed.

The turbine disk is referred to as such when in an unbladed form. When the turbine blades are installed, the disk then becomes the turbine wheel. The disk acts as an anchoring component for the turbine blades. The disk is bolted or welded to the shaft. This enables the blades to transmit to the rotor shaft the energy they extract from the exhaust gases.

The disk rim is exposed to the hot gases passing through the blades and absorbs considerable heat from these gases. In addition, the rim also absorbs heat from the turbine blades by conduction. Hence, disk rim temperatures normally are high and above the temperatures of the remote inner portion of the disk. As a result of these temperature gradients, thermal stresses are added to the stresses caused by rotation. Various means are provided to relieve, at least partially, the stresses. One way is to incorporate an auxiliary fan somewhere ahead of the disk, usually rotor shaft-driven. This will force cooling air back into the face of the disk. Another method of relieving the thermal stresses of the disk follows as incidental to blade installation. By notching the disk rims to conform with the blade root design, the disk is made to retain the turbine blades.

The turbine shaft is usually made from low-alloy steel as the material must be capable of absorbing high torque loads, such as exerted when a heavy axial flow compressor is started. The methods of connecting the shaft to the turbine disk vary. One method used is welding. The shaft is welded to the disk, which has a butt or protrusion provided for the joint. Another method is by bolting. This method requires that the shaft have a hub which matches a machined surface on the disk face. The bolts then are inserted through holes in the shaft hub. They are anchored in tapped holes in the disk. Of the two methods, the latter is more common.

The turbine shaft must have some means for joining the compressor rotor hub; this is usually accomplished by making a splined cut on the forward end of the shaft. The spline fits into a coupling device between the compressor and the turbine shafts. If a coupling is not used, the splined end of the turbine shaft fits into a splined recess in the compressor rotor hub. The centrifugal compressor engines use the splined coupling arrangement almost exclusively. Axial compressor engines may use either of these described methods.

There are various ways of attaching turbine blades. Some ways are similar to the way compressor blades are attached. The most satisfactory method used is the fir-tree design. The blades are retained in their respective grooves by a variety of methods. Some of the more common methods are peening, welding, locking tabs, and riveting. The peening method of blade retention is often used. Its use may be applied in various ways. Two of the most common applications of peening; one requires you to grind a small notch in the edge of the blade fir-tree root. You do this before installing the blade. After you have installed the blade in the disk, the notch will fill with the disk metal. The disk metal is flowed into it through a small punchmark made in the disk adjacent to the notch. The tool you use for this job is similar to a centerpunch and is usually manufactured locally. The other peening method is to construct the root of the blade so it contains the necessary retention elements. This method requires that the blade root has a stop made on one end of the root. The blade may be inserted and removed in one direction only. On the opposite end is a tang. You peen this tang over to secure the blade in the disk.

Turbine blades may be either forged or cast, depending on the metal they are made of. Turbine blades are usually machined from individual forgings. Various materials are used in the forging. Speed and operating temperatures are important factors that decide which materials go into the turbine blades. Large engines use an air-cooled blading arrangement on the GG turbine. Compressor discharge air is constantly fed through passages along the forward turbine shaft between a spacer and the shaft. A thermal shield directs the cooling air along the face of the disk for cooling of the disk. The shield is between the first- and second-stage turbine wheels. The air is then directed through slots in the fir-tree portion of the disk, into slots in the blade fir-tree. The air then goes up through holes in the blades to cool the blades.

Cooling of the turbine wheel and blades reduces thermal stresses on the rotating members. The turbine nozzles are also air cooled. By cooling the stationary and rotating parts of the turbine section, higher turbine inlet temperatures are permissible. The higher temperatures allow for more power, a more efficient engine, and longer engine life.

Power Turbines

Power turbines are used to extract the remaining energy from the hot gas. Power turbine wheels are used three different ways.

1. The aircraft jet turbine is designed so the turbine extracts only enough energy from the gases to run the compressor and accessories.

2. In the solid-wheel turbine, as much energy as possible is extracted from the gases to turn the turbine. The turbine provides power for the compressor, accessories, and the airplane propeller or the ship's generator. Examples of these engines are a turboprop airplane or ship's service generator engine. These engines are designed to run at 100 percent specified rpm all the time. The location of the mechanical connection between the turbine wheel and the reduction gear on the

compressor front shaft depends on the design of the installation. Normally, a ship's service generator cannot be disconnected from its gas turbine except by disassembly. This setup is used for generators to prevent slippage between the engine and the generator.

3. Marine propulsion engines use a combination of the two engine types. The GG has a single- or dual-stage high-pressure rotor that drives the compressor and accessories.

The PT is a multistage turbine located behind the GG turbine. There is no mechanical connection between the two turbines. The PT is connected to a reduction gear through a clutch mechanism. Either a controllable reversible pitch propeller or a reverse gear is used to change direction of the vessel.

Some ships that have two sets of engines use counter-rotating PTs. For example, PTs on one shaft rotate clockwise while the turbines on the other rotate counterclockwise. This arrangement eliminates the use of a V-drive. The GG portion rotates in the same direction for both sets of engines. The blade angle of the wheel and the nozzles in the PT section determine the directional rotation of the PT. On large ships where different length propeller shafts are permitted, the engine(s) can be mounted to the other end of the reduction gear. In this way counter-rotation of the propellers is achieved.

By varying the GG speed, the output speed of the PT can be controlled. Since only a portion of the energy is used to drive the compressor, the plant can be operated very efficiently. For example, on a cold day you can have 100 percent power turbine rpm with 80 to 90 percent gas generator rpm. These variables were discussed earlier in the chapter.

The PT is constructed much like the GG turbine. The main differences are (1) the absence of cooling air and (2) the PT blades have interlocking shroud tips for low vibration levels. Honeycomb shrouds in the turbine case-mate with the blade shrouds to provide a gas seal. They also protect the case from the high-temperature gas. Two popular methods of blade retention are the bulb type and the dovetail. These methods were discussed earlier in this chapter.

Summary

In this chapter you have learned about principles and construction of GTEs. We also discussed the evolution of the gas turbine, the theory of operation, and the classifications of the different types of engines. There are many other publications that give a more in-depth explanation of gas turbine construction. This chapter was provided to give you the basis on which to expand your knowledge of Naval gas turbine engines. You may not feel you understand the temperature-pressure relationships in a simple gas turbine at this point. If so, re-read the parts of this chapter related to theory before continuing on to the material that follows.

CHAPTER 2

THE LM2500 GAS TURBINE ENGINE

As a professional engineer and technician, you need to understand the basic construction and function of the main propulsion power plant on your ship. The LM2500 gas turbine engine has been selected as the power plant for the various class of ships in US Navy's fleet. The greater your understanding of how the engine is constructed the better you will be able to operate and maintain the engine.

In chapter 1 we discussed the basic theory of how a gas turbine engine operates and the variety of engine types available. In this chapter we discuss the LM2500 engine in particular since it is the one most often used in the US Navy.

The LM2500 is manufactured by the General Electric Company and is a marine version of the engine used in a variety of aircraft. It is the main propulsion plant for the new classes of gas turbine-powered ships. The engine is rated at approximately 20,000 brake horsepower; it has a power turbine speed of 3600 rpm on DDG-51 class. The gas turbine equipment consists of a base enclosure assembly and a gas turbine assembly. The gas turbine assembly has a gas generator, a power turbine, a high-speed flexible coupling shaft, inlet and exhaust components, and a Lube Oil Storage and Conditioning Assembly (LOSCA). The primary function of gas turbine equipment is to generate power and transmit it through a high-speed flexible coupling shaft to the ship's reduction gearbox and propeller shafting. The sections which follow describe the various components of the gas turbine equipment.

Base Enclosure Assembly

The purpose of the base enclosure assembly is to provide a thermally and acoustically insulated structure for the gas turbine assembly and connections for electrical, fire extinguishing, and air and liquid services. The base enclosure assembly consists of a shock-mounted base, an enclosure, a gas turbine mounting system, intake and exhaust systems, fire detection and extinguishing systems, a fuel enclosure heater, a lighting system and a gas turbine water wash system.

Base

The base is fabricated of steel double I-beams and steel plate to provide a stable platform for the gas turbine. Thirty-two shock mounts under the base secure the entire base enclosure assembly to the ship's foundation. These mounts lessen shock loads by absorbing most of the abrupt movements of the ship.

The gas turbine and the exhaust duct are attached to the base by 11 supports that secure the gas turbine vertically, laterally, and axially. The supports to the base attachment points are shimmed to align the gas turbine. The forward end of the gas turbine is supported by a large yoke and two additional supports attached to the front frame. The aft end is attached by four supports, three on the right side, and one on the left. The exhaust duct is secured by two supports on each side and a thrust pin underneath. The base also provides leak-tight penetration for electrical, air, CO2 or HALON, and liquid services.

NOTE: All gas turbine and enclosure references to left, right, front, rear, and clock positions apply when viewing the gas turbine from the rear (exhaust end) looking forward.

Enclosure

The enclosure is a soundproof, fire resistant housing in which the gas turbine operates. It is of double-wall construction. The inner wall is constructed of perforated metal and can withstand a temperature of 2000 °F for 15 minutes. The temperature of the outer wall normally does not exceed 150°F with ventilation air being supplied to the enclosure. The only differences between the enclosures of the ships are their access to the engine. The DDG 51 has doors on either side of the enclosure; the CG-47 has only one door with a top hatch and access ladder. In addition the CG-47 has an inlet plenum observation window.

Air Intake

The intake system supplies a high volume of air from the atmosphere to the gas turbine enclosure with minimum pressure drop. The intake ducts also provide moisture separation, silencing and anti-icing protection, and gas turbine cooling air. Blow-in doors located on the ship's upper level protect the gas turbine from air starvation in the event of inlet blockage. The intake system also allows for engine removal by a system of rails on the duct walls.

The intake section of the enclosure consists of five parts:

1. A primary inlet flexible joint which connects the ship's ducting with the enclosure. It consists of upper and lower flanges and a fiber-filled flexible boot.

2. A barrier wall which consists of four stainless steel panels bolted together. It prevents exhaust and ventilation air from being drawn into the intake. It has a removable access hatch for maintenance/operator personnel access to the inlet plenum.

3. A wire mesh inlet screen which is bolted to the barrier wall and prevents foreign objects from entering the engine.

4. An inlet duct which is bell-shaped and attached to the front frame of the compressor. The duct, or bellmouth, smooths the air flow entering the turbine. A flexible seal is attached between the inlet duct and the barrier wall.

5. A dome-shaped fairing called the center-body is attached to the compressor front frame hub to aid in smoothing the air flow.

Exhaust System

The exhaust system routes the engine exhaust gases to the atmosphere. It is designed to prevent re-ingestion of exhaust gases into the intakes and minimize heating of topside equipment. An exhaust silencer is installed in the exhaust duct to reduce the exhaust noise levels.

1. The exhaust duct is attached to the base and turbine.
2. An inner deflector is bolted to the turbine rear frame hub and protects the high-speed flexible coupling shaft from the exhaust gases.

3. An outer cone is bolted to the turbine rear frame outer flange to direct exhaust gases smoothly into the duct.

4. The exhaust extension differs in construction between the DDG-51 and CG-47 classes but serves the same purpose. It is bolted to the exhaust duct through which engine gases enter the exhaust duct. It creates an eductor effect which allows for enclosure ventilation air to exit through the space between the extension and the flexible joint.

5. The primary exhaust flexible joint connects the ship's ducting to the enclosure.

Enclosure Ventilation System

Ambient air in the enclosure ventilating system is provided by an electrically operated cooling fan. The location and operation of the cooling fan system differ among ship classes but the ducting is connected to a flexible joint common to all enclosures. Cooling air is used to maintain temperature control in the module during engine operation. Cooling air exits the module through the space around the exhaust extension. A temperature monitor is located on the enclosure ceiling just forward of the exit area. This monitor provides an alarm indication to the ship's monitoring system if the enclosure temperature exceeds a set limit. An electro-pneumatically controlled vent damper is located in the cooling air system on the enclosure overhead. It isolates the enclosure in case of fire or when the engine is secured.

Fire Detection and Extinguishing System

The detection system consists of three flame detectors, a flame detector signal conditioner, and two temperature switches. The ultraviolet flame detectors sense the presence of fire in the enclosure and generate a photo-electrical signal which is transmitted to the signal conditioner. The signal conditioner processes this input and completes the alarm control circuitry outside of the gas turbine modules, which results in an alarm indication. The temperature switches are mounted on the enclosure ceiling and generate an alarm signal if the temperature reaches a preset level.

The fire extinguishing systems differ between the DDG-51 and CG-47. On the DDG 51, when an alarm is initiated either by the detection system or by operator-initiated pushbutton located outside the enclosure, it is indicated at the ship's Local Operating Panel (LOP) and Propulsion Control Console (PCC). The operator at this point initiates casualty control procedures. These may include shutting down the engine and activating the extinguishing system. The operator activates the extinguishing system by depressing the HALON flood switch, which provides an initial HALON in the enclosure with an additional HALON available on standby.

CG-47. When the alarm is initiated either by the detection system or by operator-initiated pushbutton located outside the enclosure, a sequence of events is activated in which the engine is shut down and the fuel system is isolated from the enclosure. The cooling fan shuts down and the vent damper closes, which prevents air from entering the enclosure. The complete sequence of events is covered in another chapter. Activation of the alarm also initiates the CO2 extinguishing system. After an initial 20 second delay, CO2 discharge in the enclosure occurs. The initial discharge is 150 pounds of CO 2 and an additional 200 pounds that can be manually activated.

Both classes of ships have an inhibit switch located outside the enclosure which can disable the extinguishing when placed in the INACTIVE position.

Enclosure Heater

The heater is ceiling mounted in the enclosure and maintains the air temperature above 60 °F so suitable fuel viscosity is maintained for engine starting. It is electrically powered (440V ac) and thermostatically controlled. The heater cuts on at 60° to 70 °F and shuts off when inlet air temperature reaches 85 to 95 °F. The blower motor operates at below 125 °F and operation ceases at 145 °F. The temperature sensor is located on the enclosure ceiling near the point where enclosure air exits into the ship's exhaust.

Enclosure Illumination

Enclosure illumination is provided by explosion-proof light fixtures—several on the ceiling and base. The lights are turned on with a rotary switch mounted on the exterior wall of the enclosure near the door. The switch has four positions: off, base light, ceiling lights, and base and ceiling lights. The CG-47 is provided with two additional light fixtures in the intake plenum. These two lights are activated by pushing a button in the light switch mounted on the exterior of the forward panel.

Ice Detection

The system consists of a detector sensor located in the inlet ducting and a signal conditioner. The signal conditioner transmits a signal to the anti-icing system when icing conditions exist, which occur when the temperature is below 41 °F and humidity above 70 percent. This signal provides an alarm indicator at the control console and provides an enable signal for initiation of the anti-icing system.

Water Wash

The water wash system consists of ship's water wash/rinse supply tank and piping attached to the outside of the base enclosure. A flexible hose is attached from the inside base enclosure floor to the gas generator inlet duct at the 6-0'clock position. The inlet duct is made with an internal passageway or manifold which distributes water wash fluid to outlet spray orifices. The outlet spray orifices eject the water wash fluid into the air stream flowing through the inlet duct. The purpose of water washing is to remove contaminants from the inlet and compressor sections.

Gas Turbine Assembly

The gas turbine assembly aboard ship consists of a gas generator, a power turbine, a high-speed flexible coupling shaft, and inlet and exhaust components. The gas generator is composed of a variable geometry compressor, an annular combustor, a high pressure turbine, an accessory drive system, controls, and accessories. The power turbine is composed of a six-stage low-pressure turbine rotor, a low-pressure turbine stator, and a turbine rear frame. The high-speed flexible coupling shaft is connected to the power-turbine rotor and provides shaft power to the ship's drive system. The inlet duct and center-body are the gas turbine inlet components; the exhaust duct, outer cone, and inner deflector are the gas turbine exhaust components.

Gas Generator

The gas generator consists of a compressor, a combustor, a high-pressure turbine, and an accessory drive system. These components are discussed in detail in the following sections.

Compressor Section

The compressor is a 16-stage, high-pressure ratio, axial-flow design. Major components of the compressor are a compressor front frame, a compressor rotor, a compressor stator, and a compressor rear frame. The primary purpose of the compressor section is to compress air for combustion; however, some of the air is extracted for engine cooling and ship use. Air, taken in through the front frame, passes through successive stages of compressor rotor blades and compressor stator vanes and is compressed as it passes from stage to stage. After passing through 16 stages, the air has been compressed in the ratio of 16 to 1.

The inlet guide vanes and the first six stages of stator vanes are variable; their angular position is changed as a function of Compressor Inlet Temperature (CIT) and compressor speed. This provides stall-free operation of the compressor throughout a wide range of speed and inlet temperatures.

Compressor Front Frame. The compressor front frame provides the forward attachment point for the gas turbine, supports the forward end of the compressor section, and forms a flow path for compressor inlet air. Five struts between the hub and the outer case provide passages for lubrication oil, scavenge oil, seal pressurization air, and vent for the A sump components. The bearings of the engine are No. 3 through No. 7. The No. 3 bearing, which supports the forward end of the compressor rotor and the inlet gearbox are located in the A sump. The Compressor Inlet Total Pressure (Pt2) probe and CIT sensor are mounted in the outer case. The No. 3 strut (6 o'clock location) houses the radial drive shaft which transfers power from the inlet gearbox to the transfer gearbox mounted on the bottom of the frame.

Mounted on the forward end of the compressor front frame is a bullet nose (center body) and a inlet duct (bellmouth) assembly to direct air from the inlet plenum to the compressor. The bellmouth also contains the water wash manifold. The water wash manifold is used to inject fresh water and/or a cleaning solution into the engine while it is being motored. This is done as a maintenance procedure to clean deposits from the compressor.

Compressor Rotor. The compressor rotor is a spool/disk structure with circumferential dovetails. The use of spools makes it possible for several stages of blades to be carried on a single piece of rotor structure. There are seven major structural elements and three main bolted joints. The first-stage disk, the second-stage disk (with integral front stub shaft), and the 3- through 9-spool stage are joined by a single bolted joint at stage 2. The 3- through 9-spool stage is bolted with a stage 10 disk and the 11- through 13-spool stage is followed with the joint at stage 10. The 11- through 13-spool stage is followed by the rear shaft and a 14- through 16-spool overhung stage with a single bolted joint at stage 13. An air duct, supported by the front and rear shafts, routes stage 8 air aft through the center of the rotor for pressurization of the B sump seals. Close vane-to-rotor spool and blade-to-stator casing clearances are obtained with metal spray-rub coating. Thin squealer tips on the blades and vanes contact the sprayed material and abrasive action on the tips prevents excessive rub while obtaining minimum clearance. The first stage blades have mid-span platforms to reduce blade tip vibration.

Compressor Stator. The compressor stator consists of four sections bolted together. The front casing contains the Inlet Guide Vanes (IGV) and stages 1 through 11. The inlet guide vanes and the first six stages are variable to provide stall-free operation. The variable vanes are actuated by a pair of master levers. The aft end of the master levers are attached to pivot posts at about the 10th stage on each side of the casing. Each of the lever's forward ends is positioned by a hydraulic actuator. Fuel oil is the actuating medium. The operation of the IGVs and Variable Stator Vanes (VSV) are covered in the next chapter. The remaining vanes are stationary. The rear casing contains the 12th through the 16th stages, which are also stationary.

Three bleed manifolds are welded to the stator casings. Eighth-stage air, used for sump seal pressurization and cooling, is extracted from inside the annulus area at the tips of the hollow eighth-stage vanes. Ninth-stage air, used for power turbine cooling, power turbine forward seal pressurization, and power turbine balance piston cavity pressurization, is extracted from between the ninth-stage vanes through holes in the vane bases. Thirteenth-stage air, used for cooling the second-stage high-pressure turbine nozzle, is extracted from between the thirteenth-stage vanes through holes in the vane bases.

Compressor Rear Frame. The compressor rear frame consists of the outer case, the hub containing the B sump, and ten struts attaching the hub to the outer case. The outer case supports the combustor, the fuel manifold, thirty fuel nozzles, two spark igniters, and the first-stage high-pressure turbine nozzle support. To provide the ship's bleed air system with compressor discharge air, an internal manifold within the frame extracts air upstream of the combustion area and routes it through struts 3, 4, 8, and 9. Six borescope ports, located in the case just forward of the mid flange, permit inspection of the combustor, fuel nozzles, and the first-stage turbine nozzle.

Two borescope ports are provided in the aft portion of the case for inspection of the turbine blades and nozzles. The B sump contains the No. 4R and 4B bearings (R or no letter = Roller, B = Ball). The 4B bearing is the thrust bearing for the high-pressure rotor system. The frame struts provide passage for lube oil, scavenge oil, sump vent, seal leakage (air leakage past the compressor discharge pressure seals), and bleed air for masker, prairie, anti-icing, and engine starting services. The rear frame supports the aft end of the compressor stator by the frame's forward flange and the aft end of the compressor rotor by the No. 4R and 4B bearings.

Combustor Section. The combustor is annular and consists of four major components riveted together: cowl (diffuser assembly), dome, inner skirt, and outer skirt. The cowl assembly, in conjunction with the compressor rear frame, serves as a diffuser and distributor for compressor discharge air. It furnishes uniform airflow to the combustor throughout a large operating range. In this way, it provides uniform combustion and even temperature distribution at the turbine. The combustor is mounted in the compressor rear frame on ten equally spaced mounting pins in the forward (low temperature) section of the cowl assembly. These pins provide positive axial and radial location and assure centering of the cowl assembly in the diffuser passage.

Thirty vortex-inducing axial-swirl cups in the dome (one at each fuel nozzle tip) provide flame stabilization and mixing of the fuel and air. The interior surface of the dome is protected from the high temperature of combustion by a cooling-air film of the 16th-stage air. Accumulation of carbon on the fuel nozzle tips is minimized by venturi-shaped spools attached to the swirler.

The combustor liners are a series of over-lapping rings joined by resistance-welded and brazed joints. They are protected from the high combustion heat by circumferential film-cooling. Primary combustion and cooling air enters through closely spaced holes in each ring. These holes help to center the flame and admit the balance of the combustion air. Dilution holes are employed on the outer and inner liners for additional mixing to lower the gas temperature at the turbine inlet. Combustor/turbine nozzle air seals at the aft end of the liners prevent excessive air leakage while they provide for thermal growth.

Approximately 30 percent of the total airflow is used in the combustion process. To help explain this, it must be understood that the ideal fuel/air ratio for combustion is about 15 to 1, (i.e., 15 parts of air to one part of fuel). The rated airflow of the LM2500 engine is 123 lb per sec or 442,800 lb per hour. At rated power, the engine burns approximately 9,000 pounds of fuel per hour. Thus, at the ideal fuel/air ratio of 15 to 1, only 135,000 pounds of air per hour, 30.5 percent of 442,800 is required. The remaining 70 percent of the airflow is used for cooling, seal pressurization, and ship's service use. Approximately 5.5 percent (maximum) is used for ship's service and approximately 0.5 percent for seal pressurization. The rest is used for cooling, the majority of which reenters the mass flow cycle.

High Pressure Turbine Section. The high-pressure turbine section consists of the high pressure turbine rotor, first- and second-stage turbine nozzle assemblies, and turbine mid frame. The turbine rotor extracts energy from the gas stream to drive the compressor rotor with which it is mechanically coupled. The turbine nozzles direct the hot gas from the combustor onto the rotor blades at the optimum angle and velocity.

Turbine Rotor. The high pressure turbine rotor consists of a conical forward shaft, two discs with blades and retainers, a conical rotor spacer, a thermal shield, and a rear shaft. The forward end of the rotor is supported at the compressor rotor aft shaft by both No. 4 bearings.

Energy extracted from the hot combustion gases is transmitted to the compressor rotor through the turbine rotor forward shaft. The aft end of the turbine rotor is supported by the No. 5 bearing located in the turbine mid frame (C sump). Two seals are on the forward end of the forward shaft. The front seal helps prevent Compressor Discharge Pressure (CDP) air from entering the sump. The other seal maintains CDP in the plenum formed by the rotor and combustor. This plenum is a balance chamber that provides a corrective force that minimizes the thrust load on the No. 4B bearing. The rotor spool and both stages of blades are cooled by compressor discharge air. This air passes through holes in the first-stage nozzle support and forward turbine shaft and cools the inside of the rotor and both discs. It then passes between the paired blade dovetails and out to the blades. The first-stage blades are cooled by internal convection and external film cooling. The convection that cools the center area is done through a labyrinth within the blade. The leading edge circuit provides internal convection cooling by airflow through the labyrinth, then out through the leading edge, tip, and gill holes. Convection cooling of the trailing edge is provided by air that flows through the trailing edge exit holes. The second-stage blades are cooled by convection with all of the cooling air discharged at the blade tips. Both stages of blades are long shanked to provide thermal isolation of dovetails, high damping action for low vibration, cooling air flow paths, and low-disc rim temperature. The blades are brazed together in pairs and coated to improve erosion and oxidation resistance.

First-Stage Nozzle Assembly. The first-stage nozzle assembly consists of the nozzle support, nozzles, inner seal, outer seal, and baffles. The nozzles are coated for erosion and oxidation resistance. They are bolted to the first-stage nozzle support and receive axial support from the second-stage nozzle support. There are 33 nozzle segments in the assembly, each segment consisting of two vanes. The first-stage nozzle support forms the inner flow path wall from the compressor rear frame to the nozzle segments. Additionally it supports the nozzle segments and is bolted to the aft end of the pressure balance seal support. The nozzle assembly is air-cooled by convection and film-cooled by compressor discharge air that flows through each vane. Internally, the vane is divided into two cavities by means of two inserts. The air that flows into the forward cavity is discharged through holes in the leading edge and through gill holes located on each side near the leading edge. It then forms a thin film of cool air over the length of the vane. Air that flows into the aft cavity is discharged through trailing edge slots.

Second-stage Nozzle Assembly. The second-stage nozzle assembly consists of the nozzles, nozzle support, first- and second-stage turbine shrouds, and interstage seal. The nozzle support is a conical section with a flange that is bolted between the flanges of the compressor rear frame and the turbine mid frame. The nozzles are cast and then coated to improve erosion and oxidation resistance. There are two vanes per nozzle. The inner ends of the nozzles form a mounting ring for the interstage seal attachment. The turbine shrouds form a portion of the outer aerodynamic flow path through the turbine. They are located radially in line with the turbine blades and form a pressure seal to prevent excessive gas leakage over the blade tips. The first stage consists of 24 shroud segments; the second stage has 11 shroud segments. The interstage seal is composed of six segments bolted to the nozzles. Its purpose is to minimize gas leakage between the second-stage nozzle and the turbine rotor. The sealing surface is honeycomb and has four steps for maximum sealing. Since the honeycomb cools more rapidly than the four rotating sealing teeth, the honeycomb is pre-grooved to prevent contact under rapid or emergency shutdown conditions. The second-stage nozzle is air-cooled by convection. The nozzle vane center area and leading edge are cooled by thirteenth-stage air which enters through cooling air tubes. Some of the air is discharged through holes in the trailing edge; the remainder flows out through the bottom of the vanes and is used to cool the interstage seals and turbine blade shanks.

Turbine Mid Frame. The turbine mid frame supports the aft end of the high-pressure turbine rotor (No. 5 bearing); it also supports the forward end of the low-pressure turbine rotor (No. 6 bearing). It is bolted between the aft flange of the compressor rear frame and the forward flange of the power turbine stator. The frame provides smooth diffuser flow passage for high-pressure turbine exhaust gas into the power turbine. The frame hub is an open, drum shaped, one-piece casting with flanges to support the C sump housing, stationary seals, inner liner support, and power turbine first-stage nozzle support. The C sump contains the No. 5 and No. 6 bearings. Eight struts connect the hub to thc outer case. The struts provide passage for C sump lubrication and scavenge oil, cooling air, sump vent, and seal drain services.

The outer case contains ports for high-pressure turbine exhaust gas thermocouples and pressure probes. The ports also provide access for borescope inspection of the power turbine inlet area. The frame liner assembly consists of an inner and outer liner held together with airfoil-shaped strut fairings butt-welded to both liners. This assembly guides the gas flow and shields the main structure from a high temperature. The power turbine first-stage nozzle assembly is part of the turbine mid frame assembly. This assembly consists of fourteen segments of six vanes per segment. The inner end is bolted to the nozzle support; the outer end is secured between the mid frame aft flange and the power turbine stator forward flange.

Accessory Drive Section. The accessory drive section consists of an inlet gearbox in the hub of the front frame, a radial drive shaft inside the 6 o'clock strut of the front frame, and a transfer gearbox bolted underneath the front frame. The fuel pump and main fuel control, the pneumatic starter, and the lube and scavenge pump are mounted on the aft side of the transfer gearbox. An air-oil separator on the front is a part of the gearbox.

Power to drive the accessories is extracted from the compressor rotor through a large-diameter hollow shaft which is spline-connected to the rotor front shaft. A set of bevel gears in the inlet gearbox transfers this power to the radial drive shaft. This transmits the power to another set of bevel gears in the forward section of the transfer gearbox. A short horizontal drive shaft transmits the power to the accessory drive adapters in the transfer gearbox.

Inlet Gearbox. The inlet gearbox assembly consists of a cast aluminum casing, a shaft, a pair of bevel gears, bearings, and oil jets. The casing, which is bolted inside the front frame hub, mounts two duplex ball bearings and a roller bearing. It has internal oil passages and jets to provide lubrication for the gears and bearings. The shaft, which rotates on a horizontal axis, is splined at the aft end to mate with the second-stage disk of the compressor rotor. The forward end of the shaft mounts the upper bevel gear and is supported by a duplex ballbearing. The lower bevel gear rotates on a vertical axis. It is supported it its upper end by a roller bearing and at its lower end by a duplex ballbearing. The lower end is also splined to mate with the radial drive shaft.

Radial Drive Shaft. The radial drive shaft, a hollow shaft externally splined on each end, mates with the bevel gears in the inlet and transfer gearboxes. Its function is to transmit power from the inlet gearbox to the forward section (bevel gearbox) of the transfer gearbox. The shaft contains a shear section to prevent damage to the accessory drive system.

Transfer Gearbox. The transfer gear-box assembly consists of a two-piece aluminum casing, an air-oil separator, gears, bearings, seals, oil nozzles, and accessory adapters. The forward section (bevel gearbox) contains a set of right-angle bevel gears and a horizontal drive shaft which transmits the power to the gear train in the rear section (accessory gearbox). Each bevel gear is supported by a duplex ballbearing and a roller bearing. An access cover in the bottom of the casing facilitates installation of the radial drive shaft. The "plug-in" gear concept is used on all accessory adapters and idler gears in the aft section. This permits an entire gear, bearing, seal, and adapter assembly to be removed and replaced without disassembling the gearbox. Each spur gear is supported by a casing-mounted roller bearing on one end and an adapter-mounted ballbearing on the other end. The accessory drive spur gears are internally splined. Internal tubes and oil nozzles provide lubrication of the gears and bearings. Gearbox carbon-face seals are retained from the outside of the gearbox and can be replaced without disassembly of the gearbox.

Gearbox Mounted Accessories. Mounted on the aft side of the transfer gearbox are four major accessories: the pneumatic starter, the lube and scavenge pump, the fuel pump and the main fuel control. The main fuel control is mounted on the aft side of the fuel pump.

The air-oil separator, starter, lube scavenge pump, fuel pump, main fuel pump operation, and functions are covered in further detail in the following chapter.

Power Turbine Section

The power turbine section consists of a six-stage low pressure turbine rotor, a turbine stator, and a rear frame.

Turbine Rotor. The power turbine rotor consists of six discs bolted together to form the rotor spool. Blades of all six stages contain interlocking tip shrouds for low vibration levels and are retained in the discs by dovetails. Replaceable rotating seals, secured between the disc spacers, mate with stationary seals to prevent excessive gas leakage between stages. The first three stages of blades are coated for corrosion protection.

Turbine Stator. The power turbine stator consists of two casing halves, stages two through six turbine nozzles, and six stages of blade shrouds. The first-stage nozzle is part of the turbine mid frame. Honeycomb shrouds, mounted in casing channels, mate with the shrouded blade tips to provide close-clearance seals. The stationary interstage seals are attached to the inner ends of the nozzle vanes to maintain low leakage between stages. Insulation is installed between nozzle shrouds and casing to protect the casing from the high temperature of the gas stream.

Turbine Rear Frame. The turbine rear frame consists of an outer case, a single-piece cast steel hub, and eight equally spaced radial struts connecting the case and the hub. It forms the exhaust gas flow path and supports the aft end of the power turbine rotor (No. 7B and 7R bearings) and the forward end of the high-speed flexible-coupling shaft. The hub contains the D sump; the D sump contains the No. 7B and 7R bearings (B = Ball, R = Roller). The No. 7B bearing is the thrust bearing for the power turbine. The struts provide passage for lubrication and scavenge oil, C and D sump seal pressurization air, D sump vent, cooling air, and power turbine balance piston air services. The two power turbine speed pickups also pass through the struts (No. 3 and No. 7).

High-Speed Flexible-Coupling Shaft

The high-speed flexible-coupling shaft consists of a forward adapter which mates with the power turbine, two flexible couplings, a distance piece, and an aft adapter which mates with the reduction gear high-speed pinion. The forward and aft adapters are connected to the distance piece by the flexible couplings. The flexible couplings allow for axial and radial deflection between the gas turbine and the ship's drive system during operation. Inside the aft adapter and the aft flexible coupling is an axial damper system consisting of a cylinder and piston assembly. The damper assembly prevents excessive cycling of the flexible couplings. Anti-deflection rings restrict radial deflection of the couplings during shock loads.

Lube Oil Storage Conditioning Assembly (LOSCA)

The LOSCA is remotely located from the gas turbine and consists of an oil storage tank, a heat exchanger, a duplex filter and a scavenge oil check valve. Its function is to provide and filter the synthetic lube oil used to lubricate the gas turbine engine. The LOSCA is an integral part of the engine. Its function and operation are covered later as part of the lube oil system.

Summary

This chapter has covered the parts of the LM2500 gas turbine engine found aboard the various classes of ships. The base enclosure is shock mounted in the engine room and contains the gas turbine engine. It is a connecting point for the inlet and exhaust ducting. It provides penetration points for piping and connection points for electrical connections.

The engine itself consists of a gas generator, a high-pressure turbine, and a power turbine. The gas generator compresses atmospheric air at a 16 to 1 ratio and mixes it with fuel. Once ignited, the combustion gases turn the high-pressure turbine. When the engine is at a self-sustaining speed, the high-pressure turbine drives the compressor. The combustion gases also drive the power turbine which is connected to the reduction gear by a high-speed flexible coupling. The accessory gearbox provides the driving force for the fuel pump and main fuel control—the lube oil pump, air-oil separator, and a mounting pad for the starter. With the knowledge you have gained from this chapter, you should more easily be able to follow the actual operation of the engine and its many components.

CHAPTER 3

LM2500 GAS TURBINE ENGINE OPERATION

As an engineer, you need to understand the operation of the gas turbine engine. In chapter 2 we discussed the construction of the engine and what function each component performed. In this chapter we discuss the actual operation of the engine. How the engine is started, what occurs within the engine and its systems while it is running, and the various operating parameters are covered. The control systems of the various ships differ and are discussed fully in other publications. This chapter is confined strictly to the operation of the engine common to all classes of ships.

It's important to fully understand the function of the engine systems and controls in order to recognize and troubleshoot problems that may occur during engine operation. We will discuss the engine operation with the initiation of a start and follow it through the flow of air. We describe the operation of the ignition system, the fuel system, and how the main fuel control and Power Level Angle (PLA) actuator functions in relation to the inputs and outputs of the Free Standing Electronics Enclosure (FSEE). We also discuss more fully the lubrication system which we touched on briefly at the end of chapter 2.

Gas Turbine Start-Air System

Start air is used for two separate purposes on the gas turbine engine. First, it is used to turn the compressor and high-pressure turbine through the starter and accessory gear box until the engine reaches self-sustaining speed. Second, it is used for motoring the gas turbine engine during water washing procedures and other evolutions, such as purging the engine of fuel prior to start.

The sources of start air differ with the various classes of ships. On the CG-47 and DDG-51 classes, bleed air from the ship's bleed air main is normally used for starting. This air is provided by ship's gas turbine generators (501K-17/K-34) and main engines (LM2500). The CG-47 and DDG-51 classes also have an emergency starting air capability from the ship's 3000-psi high-pressure air system through a pressure-reducing valve and orifice that reduce the pressure to 85 psi.

Starter Air

Starter air is cooled to approximately 400 °F for starting. The bleed air on the DD and CG classes of ships is cooled by mixing hot bleed air with cool masker air through a mixing valve (discussed in more detail when the bleed air system is presented). On the DDG class ship, bleed air is cooled by a starting air cooler which uses seawater as the cooling medium and a start air temperature regulating valve. Some hot air bypasses the cooler and uses the temperature regulating valve to maintain a constant air temperature of 400 °F. The air pressure is regulated at 45 psi for starting and 22 psi for motoring.

The following description of operation for the regulating valve covers the CG classes of ships. However, it is basically the same for the DDG class except the motoring function actuates manually instead of electrically.

GTM Motor Air Regulating Valve

A GTM motor air regulating valve is located prior to the point where the start/motor air piping branches to both GTMs. Each GTM motor air regulating valve is a 4-inch, pneumatic piston-actuated, butterfly-vane, solenoid- controlled, pressure-regulating valve. Each regulates air at 45 psig to 20 to 22 psig for GTM motoring. The GTM motor air-regulating valve consists of a regulator assembly and an actuator assembly. The regulator assembly consists of the valve body and butterfly vane. The actuator assembly consists of the solenoid operated pilot valve with manual override, downstream pressure-sensing diaphragm-operated poppet valve assembly, and the actuating control piston and control linkage. A set of spring-loaded micro-switches is also provided to indicate valve status.

Inlet air pressure from the start air filter enters the valve body where pressure is ported to the pilot valve. When the solenoid deenergizes, inlet air pressure from the valve body shuts off at the pilot valve and the underside of the actuating control piston vents to the atmosphere. This causes the actuating control piston to remain in the down position, which keeps the butterfly vane in the full open (non-regulated) position thereby having no effect on the starting air pressure. When the solenoid energizes, the pilot valve vent closes and inlet air pressure from the valve body emits to both sides of actuating control piston through fixed-rate chamber orifices to drive the actuating control piston. This rotates the butterfly vane and regulates the downstream air pressure. The position of the actuating control piston is determined by the diaphragm operated, poppet-valve assembly. This assembly opens and closes the poppet valve to control the venting of the chamber area above the actuating control piston in relationship to the downstream sensed static air pressure. As downstream sensed static air pressure increases, the poppet-valve flow opening increases. This allows air to be vented from the chamber area above the actuating control piston. This, in turn, allows the piston to move up because the air pressure below the piston is greater than the spring force and air pressure above the piston. As the actuating control piston moves up, the butterfly vane through control linkage rotates toward the closed position and thereby reduces downstream air pressure. As downstream air pressure decreases, the reverse action takes place until the airflow into and out of the upper chamber is proportionally balanced. At that point, the actuating control piston position is maintained as is downstream pressure until there is a change in the downstream pressure due to an increase or decrease in flow.

Starter Air Valve

The starter valve is a normally closed pneumatic regulator and shutoff valve. The valve contains both electrical and mechanical position indicators and is both pneumatic and spring loaded in the closed position. Air from the ship's start air system is supplied through an inlet fitting on the enclosure base to the starter valve at 0 to 75 psig. When 28V d.c. power is supplied to the solenoid, the valve opens regulating discharge air pressure (to the starter) at 35 to 41 psig. It consists of a bleed-on regulator, a solenoid switcher, a pneumatic switcher, a check valve, an actuator, and a butterfly valve.

When the solenoid energizes, the ambient vent of the solenoid switcher closes and regulated pressure ports to the piston of the pneumatic switcher. The pressure passes through the check valve to the opening chamber of the actuator. The piston in the pneumatic switcher allows pressure in the closing chamber of the actuator to bleed to the outlet side of the valve. Pressure in the opening chamber increases as pressure in the closing chamber decreases. This causes the butterfly valve to open.

When inlet pressure is lower than the desired outlet pressure, forces acting on the actuator are such that the butterfly is held in the full open position. When outlet pressure increases to the desired level, the actuator forces balance and the butterfly moves to an intermediate position. Any slight change in outlet pressure causes a corresponding modulation of the butterfly position to maintain the desired outlet pressure.

When the solenoid switcher deenergizes, it shuts off regulated pressure to the pneumatic switcher and to the closing chamber of the actuator, at the same time opening the pneumatic switcher and venting the closing chamber to ambient. The pneumatic switcher ports regulated pressure to the closing chamber of the actuator and shuts off the outlet bleed of the closing chamber. The check valve opens to permit full flow of closing chamber air to ambient to effect fast closing of the butterfly valve.

Starter

The pneumatic starter consists of an inlet assembly, a turbine assembly, reduction gearing, a cutout switch, an overrunning clutch, and a splined output shaft. The turbine is a single-stage, axial-flow type. The reduction gearing is a compound planetary system with a rotating ring gear. The overrunning clutch is a pawl and ratchet type, which provides positive engagement during starting and overrunning when driven by the gas generator. The cutout switch is normally closed and is actuated by a centrifugal governor which trips the switch. The output shaft has a shear section to prevent overtorque damage. Compressed air supplied to the inlet stator housing is directed to the turbine wheel through the stator nozzle. A gear-set transforms the high speed and low torque of the turbine wheel to low speed and high torque of the output shaft. Exhaust air is released through the perforated exhaust screen at approximately ambient pressure. The drive shaft assembly disengages from the output shaft after the gas generator speed exceeds the drive-shaft speed. This allows the output shaft to rotate with the gas generator while the driveshaft and gears coast to a stop.

Starter air is approximately 400 °F and regulated at 35 to 40 psi for start or 20 to 22 psi for motoring. In actual operation, starter air is directed to the starter turbine blades. The starter is mounted on the transfer gear box. The gear box is part of the accessory drive section. Turbine high-speed low torque is transformed into low-speed high torque through 16 to 1 ratio reduction gears. The starter rotates the gears in the transfer gear box which in turn rotates the radial drive shaft. The radial drive shaft is mated with the inlet gear box; it in turn, mates with the second stage disc of the compressor. The compressor is brought up to speed and a centrifugal switch cuts out the air to the starter at approximately 4300 to 4700 rpm. The starter drive shaft disengages from the output shaft.

The turbine wheel reaches speeds up to 75,000 rpm. The starter is lubricated from a self-contained sump that uses synthetic lube oil (MILSPEC 23699). The time the starter can be operated is limited and starter duty cycles for starting and motoring can be found in the ship's EOSS. Refer to EOSS for the proper cycles.

Primary Air Flow

The gas generator draws air from the intake duct, through the enclosure inlet, plenum, inlet screen, bellmouth, and front frame. After the air is compressed in the 16th stage compressor, it enters the combustion section where some of it mixes with the fuel and burns. The remainder of the air is used to center the flame in the combustor and for cooling the combustor, the high-pressure turbine rotor and blades, and the first-stage high- pressure turbine nozzle. Hot gas from the combustor passes through the high-pressure turbine where it extracts some of the energy by the high-pressure turbine rotor and uses it to turn the compressor. The hot gas leaves the high-pressure turbine and passes through the turbine mid frame; it then enters the power turbine where the power turbine rotor extracts most of the remaining energy. This drives the high-speed flexible-coupling shaft. The shaft provides the power to the reduction gear high-speed pinion. The gas leaves the power turbine, passes through the turbine rear frame into the exhaust duct, and out through the uptake duct.

Bleed Air Flow

Air extracted from the compressor for cooling, seal pressurization, pressure balance to reduce rotor thrust loads, and for ship's service is called bleed air. In this installation, air is bled from stages 8, 9, 13, and the 16th stage compressor discharge pressure (CDP). Only sixteenth-stage air is used for the ship's bleed air system.

Eighth-Stage Air

Eighth-stage air is bled from the compressor through hollow eighth-stage stator vanes into an external manifold. From the manifold, the air is piped forward and aft to ejector nozzles. Each ejector contains a venturi through which the eighth-stage air passes; it draws enclosure air into the ejector. This air mixes with the eighth-stage air, reduces the downstream pressure and temperature and increases the volume. Air from the forward ejector is piped into the front frame hub; there it pressurizes and cools the A sump. Some of the A sump air passes through holes in the compressor rotor front shaft, through the rotor air duct, and through holes in the rotor aft shaft; there it pressurizes and cools the B sump. Air from the rear ejector is piped into the turbine rear frame hub; it is used to pressurize and cool the D sump. Part of the air entering the D sump bleeds into the flexible- coupling shaft tunnel for cooling and passes out the aft end of the exhaust duct. Some of the D sump air passes through holes in the power turbine aft shaft, rotor air tube, and holes in the rotor front shaft; there it pressurizes and cools the C sump.

Ninth-Stage Air

Ninth-stage air is bled from the compressor through holes in the ninth-stage vane bases and compressor casing into an external manifold, where it is piped to the turbine mid frame and turbine rear frame. Air enters the turbine mid frame through all the struts. Some of the air exits through holes in the frame hub to cool the frame inner liner. The rest of the air enters tubes in the C sump air seals; after it crosses these seals, the air passes through and cools the power turbine rotor and then exits into the power turbine exhaust gas. The air to the turbine rear frame enters the frame through struts 2 and 8 and passes into an area between, where it acts as a balance chamber to reduce the aft loading on the No. 7B bearing.

Thirteenth-Stage Air

Thirteenth-stage air is bled from the compressor through holes in the thirteenth-stage vane bases and compressor casing into an external manifold. The air is piped through the compressor rear frame casing and into the high-pressure turbine shrouds. The air then flows through and cools the second-stage turbine nozzle. Some of the air exits through nozzle-trailing edge holes. The remaining air is used for cooling the interstage seal, the aft side of the first-stage blade shanks, and the front side of the second-stage shanks.

Sixteenth-Stage Air

Sixteenth-stage, compressor discharge bleed air is bled through holes in the inner wall of the compressor rear frame and out through frame struts 3, 4, 8, and 9 and piped to the bleed air valve. The bleed air valve operates electrically to provide air to the ship's bleed air system. High-pressure turbine rotor and blade cooling air is extracted internally through the aft stationary air seal and holes in the forward end of the rotor front shaft. The remaining compressor discharge bleed air is used for cooling the combustion liner and first-stage high-pressure turbine nozzle vanes.

Balance Piston Air

Balance piston air is bled into a chamber to apply force to the face of a rotor disc in the direction opposite the thrust load in order to reduce the axial load applied to a thrust bearing by the rotor thrust. This increases the thrust-bearing life and is known as balance piston air or pressure balance air.

Fuel and Speed-Governing System

The fuel and speed-governing system regulates and distributes fuel to the combustion section of the gas generator to control gas generator speed. The power turbine speed is not directly controlled, but is established by the gas stream energy level produced by the gas generator. Power turbine overspeed protection is provided by an electronic overspeed switch that is located in the Free Standing Electronic Enclosure (FSEE) via signals from the two power turbine speed pickups. The fuel and speed-governing system consists of a fuel pump and filter, the Main Fuel Control (MFC), the fuel shutdown valves, the fuel nozzles, and igniters.

To assure an adequate supply of fuel for gas turbine operation, the fuel pump has a higher flow capacity than the gas turbine uses. Within the control the fuel is divided into metered flow and bypass flow. This division maintains a preset pressure drop across the metering valve by the use of a bypass valve. Bypass fuel is ported to the high-pressure element inlet screen of the fuel pump. If an abnormal condition occurs that causes pump outlet pressure to become too high, a relief valve in the pump bypasses fuel back to the high-pressure element inlet screen.

A pressurizing valve, mounted on the fuel control outlet port, maintains back-pressure to ensure adequate fuel pressure for control servo operation. Two electrically operated fuel shutdown valves connected in series provide a positive fuel shutoff. When the fuel shutdown valves are open, metered fuel for gas turbine operation flows from the fuel control, through the pressurizing valve, shutdown valves, fuel manifold, and fuel nozzles. When the fuel shutdown valves are closed, metered fuel is bypassed to the fuel pump inlet; the fuel drain ports in the valves open to allow fuel remaining in the manifold, nozzles, and lines to drain. Thirty fuel nozzles, which project through the compressor rear frame into the combustor, produce an effective spray pattern from start to full power.

The fuel and speed governing system controls the variable stator vanes to maintain satisfactory compressor performance over a wide range of operating conditions. At high inlet air temperature and low compressor speed, the larger forward stages of the compressor are capable of pumping more air than the smaller aft stages. Because of this characteristic, the aft stages become over-loaded. This causes the airflow to stop and possibly reverse. This is known as compressor stall. It is prevented by having the Inlet Guide Vanes (IGV) and first six stages of stator vanes variable. The fuel and speed governing system controls the variable vanes, scheduling them toward the closed position when compressor speed drops or inlet temperature rises; thereby matching the output of the forward stages to that of the rear stages.

Fuel Pump and Filter

Fuel supplied by the fuel oil service system flows through the base inlet connector of the fuel pump and filter. The fuel pump contains two pumping elements, a centrifugal boost element and a high-pressure gear element. It provides mounting pads and flange ports for the fuel filter and the main fuel control. This feature reduces the amount of external piping required. The pump also provides a drive shaft for the main fuel control. This eliminates the need for a separate transfer gearbox drive pad.

Fuel from the ship's supply enters the pump through the fuel inlet port and is boosted in pressure by the centrifugal boost element, discharging into a circumferential scroll. The flow passes through a screen which has an integral bypass; it then passes into the high-pressure positive displacement gear element. The combination of pumping elements is designed to provide improved fuel pump features so that normal operation can be sustained without external boost pumps. The pump incorporates a high-pressure relief valve. This valve cracks at or above 1350 pounds per square inch absolute (psia); it reseats at or above 1325 psia. These features protect the pump and downstream components against excessive system pressures.

The fuel filter is a high-pressure filter mounted on the fuel pump and flange-ported to eliminate external piping. The head houses a bypass relief valve; the bowl houses the filter element. The filter element is rated at 46 microns nominal and 74 microns maximum. It prevents larger contaminants from being carried into the main fuel control.

High pressure fuel flows from the fuel pump through the flange port and enters the filter bowl. The fuel then flows from the outside of the filter element to the center, up into the head, out the flange return port, and back into the fuel pump. There, it is routed to the main fuel control. If the filter becomes clogged, the bypass relief valve opens at 35 pounds per square inch differential (psid).

Main Fuel Control (MFC) and Pressurizing Valve

The main fuel control is basically a speed governor which senses gas generator speed and power lever position; it adjusts the fuel flow as necessary to maintain the desired speed set by the power lever. The control is a hydro-mechanical device which operates by use of fuel-operated servo valves. It performs the following functions:

1. Controls speed by metering fuel to the fuel nozzles during acceleration, deceleration, and steady-state operation. Excess fuel supplied by the fuel pump to the control is returned to the pump downstream of its low-pressure element. The control also uses the fuel from the pump as a hydraulic medium.

2. Alters the fuel schedule automatically to maintain the speed setting and establishes fuel limits for acceleration and deceleration. Three fuel schedules are established by the control: acceleration, deceleration, and minimum fuel schedules. The acceleration schedule limits fuel flow necessary for acceleration to prevent over-temperature and stall. The deceleration schedule limits the rate of fuel flow decrease to prevent combustion flameout during deceleration. The minimum fuel schedule limits fuel flow for starting to prevent over-temperature. The gas turbine parameters vary, so the fuel limits vary to provide optimum acceleration and deceleration schedules. In order for the

control to determine the schedules, certain parameters must be sensed. These parameters are Compressor Discharge Pressure (CDP), Compressor Inlet Temperature (CIT), and Gas Generator Speed (N_{GG}). The control, using hydro-mechanical mechanisms, senses the parameters and computes a limit. The computed limit is compared with actual fuel flow and controls the metering valve should the governor attempt to exceed the limit.

The MFC schedules the variable stator vanes as a function of gas generator speed and CIT. Actual position of the variable stator vanes is sensed by the control via a position feedback cable. One end of the feedback cable is connected to the left master lever arm; the other end is connected to the feedback lever on the MFC. The control may be divided into a number of working sections, each of which serves a specific function. The operation of these sections is described in the following paragraphs.

Fuel Supply. Fuel is supplied to the control at fuel pump discharge pressure. The pump supplies more fuel than is required for any engine operating condition. Required fuel flow is determined by a metering valve orifice area. Excess fuel is directed through a bypass valve to the fuel pump. Regulated pressure in the case of the control is used as a reference pressure in the servo-mechanisms.

One of the three parameters which determine the fuel flow limits is gas generator speed. The fuel control contains a three dimensional (3-D) cam. It is rotated in proportion to gas generator speed.

The second of the three parameters is Compressor Inlet Temperature (CIT). The three-dimensional cam is moved axially in relation to compressor inlet temperature.

The third parameter is Compressor Discharge Pressure (CDP). Changes in CDP rotate the CDP cam which is mechanically linked to the 3-D cam by levers which combine the output of both cams in adding linkage.

For each gas generator speed and CIT combination, there is a unique position of the 3-D cam. The combination of the 3-D cam position and the CDP cam output in the adding linkage define the allowable acceleration fuel flow limit. A third input represents metered fuel input to the gas generator. This input is combined to the adding linkage to determine acceleration and deceleration limits. Power demand is applied through the Power Lever Angle (PLA) actuator to a speed governor. The speed governor senses gas generator speed. The governor automatically controls flow of fuel to the gas generator to maintain a constant gas generator speed. This is done by means of a flyweight and pilot valve to ensure that the gas generator never exceeds a predetermined safe operating speed, fuel flow is reduced by means of a force applied to a constant differential pressure valve.

The following external adjustments are provided on the control:

1. *Fuel Specific Gravity.* This affects minor adjustments of minimum and acceleration fuel flow to compensate for specific gravity differences of the various types of fuels. It is adjusted by aligning a mark on the adjusting screw head with the specific gravity marked on the adjustment plate. The plate corresponds to the measured specific gravity of the fuel being used. The specific gravity setting should always match the measured specific gravity of the fuel being used.

2. *Idle Speed.* This adjusts minimum steady state (idle) speed. Clockwise adjustment of the adjusting screw increases speed; counter- clockwise adjustment decreases speed. Speed change is approximately 25 rpm per click. There are 36 clicks per revolution of the adjusting screw. An adjustment is required if idle speed is not between 4900 and 5000 rpm.

3. *Power Trim*. This adjusts maximum steady state speed. However, for this application it is not required. The maximum speed setting is 9827 rpm. The speed is set by the governor manufacturer and exceeds the maximum speed requirements for this application.

4. Stator Position Feedback Cable. This is a trim adjustment to open or close the variable stator vanes to the nominal stator schedule. Clockwise adjustment of the adjusting screw closes the stators; counterclockwise adjustment opens the stators. Stator change is 0.7 degrees per turn of the adjusting screw.

Minimum speed (idle) adjustment and stator position feedback cable adjustment are normally made only during engine acceptance, testing, or a checkout run. A checkout run follows a component replacement which affects any of these parameters, such as replacement of the main fuel control.

Pressurizing Valve. The pressurizing valve pressurizes the fuel system to provide adequate fuel control servo supply pressure and variable stator vane actuation pressure. This is necessary for proper fuel and stator vane scheduling during gas generator operation at low fuel flow levels. The valve is a fuel pressure-operated, piston valve. The piston is held on its seat (closed) by spring force and fuel pressure (reference pressure) from the main fuel control. Servo pressure is 110 to 275 pounds per square inch gage (psig). Main fuel control discharge fuel (metered fuel for combustion) enters the pressurizing valve at the opposite side of the piston. When main fuel control discharge pressure is 80 to 130 psig greater than reference pressure, the valve opens. Thus, the upstream pressure (including servo supply and stator actuation) is 190 psig or greater before the pressurizing valve opens, which is adequate for proper operation.

Variable Stator Vanes

The variable vanes are positioned by two hydraulic actuators operated by fuel pressure from the MFC. The MFC has a variable vane scheduling cam positioned by engine speed and Compressor Inlet Temperature (CIT) signals; it has a variable vane feedback mechanism which receives a vane position signal from a linkage connected to the variable stator vane master lever; it also has a variable vane pilot valve positioned as a result of the comparison of the scheduling cam position and the feedback signal. Changes in engine speed rotate the scheduling cam; changes in CIT translate the cam. Movement of the cam repositions the pilot valve. The pilot valve ports high-pressure fuel (pump discharge pressure) to either the rod end (closing) or head end (opening) of the vane actuators; it vents the other end to bypass pressure. The variable vane actuating linkage mechanically transmits the actuator movement to the variable vanes and IGV. A flexible cable is attached to the linkage. It transmits a feedback signal to the MFC. The feedback mechanism in the MFC repositions the pilot valve to terminate the actuator signal when the vanes reach the scheduled position.

Variable Stator Vanes (VSV) Actuators are single-ended, uncushioned hydraulic cylinders which are driven in either direction by high pressure fuel. The piston stroke is controlled by internal stops. The actuators are mounted tangentially on the compressor stator forward flange at the 3 o'clock and 9 o'clock positions. They are connected to the variable stator vanes through master lever arms and actuation rings. The MFC schedules high-pressure fuel to either the head-end port (opens VSV) or the rod-end port (closes VSV). Control parameters sensed by the MFC to schedule variable vane angle are Gas Generator Speed (N_{GG}) Compressor Inlet Temperature (CIT), and stator vane angle via a feedback cable. The feedback cable is connected on one end to the left master lever arm and on the other end to the MFC.

Fuel Shutdown Valves

Two identical, pilot-valve-actuated, electrically controlled fuel shutdown valves are used on the gas generator. When energized, they port the metered fuel to the gas generator fuel manifold. When de-energized, they port the fuel to bypass (at the fuel pump inlet). The fuel shutdown valves are in series in the fuel lines; they are in parallel electrically. Thus both valves must be energized to permit fuel flow to the manifold. De-energizing of either valve shuts off fuel flow to the manifold. In operation, pilot pressure is supplied to the valves by the fuel pump.

When the valves are energized, internal pilot valves port this pressure to actuate the main shutdown valves. When de-energized, the pilot valves cause the shutdown valves to port the inlet pressure to the fuel pump inlet and the manifold to drain. Each shutdown valve has a pilot pressure bleed relief valve to prevent backflow through the valve during gas turbine motoring. A check valve which has a 2 psid cracking pressure is located in the bypass port of the No. 2 shutdown valve to prevent fuel backflow into the No. 2 valve from the No. 1 valve bypass line.

Purge Valve

The purge valve operates electrically. It is a normally closed, on-off valve used to drain low temperature fuel from the system prior to gas turbine start. It is spring-loaded to the closed position; a solenoid opens the valve when energized. Approximately 3 gallons of fuel are drained from the system during purging.

Fuel Nozzle

The fuel nozzle is a dual orifice, swirl atomizer with an internal flow divider. Thirty fuel nozzles produce the desired spray pattern over the full range of fuel flows. Fuel enters the nozzle through a single tube, flows through a 117-micron screen, and then the flow divider. When the nozzle is pressurized, primary fuel flows through a drilled passage and tube assembly in the nozzle shank. It then flows through the primary spin chamber and into the combustor. When fuel pressure to the nozzle rises to 330 to 350 psig, the flow divider opens and introduces a secondary fuel flow. The secondary fuel flows through the flow divider, through a passage in the nozzle shank, into the secondary spin chamber, and mixes with the primary flow as it enters the combustor. An air shroud around the nozzle tip scoops a small quantity of air from the main airstream to cool the nozzle tip. This retards the buildup of carbon deposits on its face.

Ignition System

The ignition system consists of ignition exciters and spark igniters. The ignition system is actuated by an electronic signal during the engine start sequence. It can be generated either manually by the operator or by the electronic timing controls. The engine must reach a speed of 1200 rpm prior to ignition; then, if operated manually the igniters should energize prior to the fuel valve. When you start in the automatic mode, the fuel valves do not open unless the igniters are already functioning. This prevents excess fuel from entering the combustion chamber and causing excessive start temperatures. The igniters are secured by an electronic signal at 4500 rpm. By this time combustion occurs and the engine reaches self-sustaining speed.

The ignition exciters are the capacitor discharge type. They are on the right side of the front frame and are attached to special mounts which absorb shock and vibration. The exciters operate on 115V, 60 Hz input. The power transforms, rectifies, and discharges in the form of capacitor discharge energy pulses through the coaxial-shielded leads to the spark igniters.

Spark igniters are of the surface gap-type. They have internal passages for air-cooling and air vents to prevent the accumulation of carbon in interior passages. The igniter has a seating flange with attached copper gaskets for sealing purposes. Grooves are cut in the outer surface of the tip and axial holes for cooling the outer and inner electrodes with compressor bleed air.

The ignition leads are low-loss connections between the ignition exciters and the spark igniters. They have metallic shielding which incorporates copper inner braid, sealed flexible conduit, and nickel outer braid (coaxial).

Lube Oil System

The lubrication system provides the gas turbine bearings, gears, and splines with adequate cool oil to prevent excessive friction and heat. The synthetic lube oil used in this application is MIL-L-23699. The lubrication system is a dry sump system divided into three subsystems identified as lube supply, lube scavenge, and sump vent. Lube oil is gravity fed from the Lube Storage and Conditioning Assembly (LOSCA), through the ship's piping to an inlet fitting in the enclosure base, then to the inlet of the supply element of the lube and scavenge pump. From the supply element of the pump, the oil passes through the supply duplex filter, through a check valve, and into a supply manifold. From the supply manifold, the oil is distributed to the four sumps and the transfer gearbox.

The seven principal bearings that support the two separate rotating systems are located in four sump enclosures as follows:

- Bearing No. 3 in the A sump (Roller)
- Bearing No. 4R and 4B in B sump
- Bearing No. 5 and 6 in C sump (Roller)
- Bearing No. 7B and 7R in D sump (R-Roller Bearing, B-Ball Bearing)

Oil from the lube supply system is fed to each of the sumps under controlled pressure and directed by jets on to the bearings. Each of the sumps are encased in protective air jackets, maintained by air seals. This prevents excessive heat from reaching the oil-wetted walls of the inner oil seals which prevents coking and thermal deterioration of the oil. Each end of the sump has a labyrinth/windback oil seal and a labyrinth air seal which prevents oil leakage from the sumps.

The oil seals are of two types: labyrinth/windback used in the sump areas and carbon seals used in the transfer gearbox. The labyrinth/windback seals combine a rotating seal which has oil slingers and a serrated surface with a stationary seal which has windback threads and a smooth rub surface. The oil slingers throw oil into the windback threads which direct the oil back to the sump area. The serrations cut grooves into the smooth surface of the stationary seal to maintain close tolerances throughout a large temperature range. This seal allows a small amount of seal pressurization air to leak into the sump; this prevents oil leakage. The carbon seal consists of a stationary spring-loaded carbon sealing ring and a rotating, highly-polished steel mating ring. This arrangement prevents oil in the gearbox from leaking past the drive shafts of the starter, fuel pump, and auxiliary drive pad.

The gas turbine air seals are of two types: labyrinth/honeycomb (used in the sump and turbine areas) and fishmouth (used in the combustor and turbine mid-frame). The labyrinth/honeycomb seal combines a rotating seal that has a serrated surface with a stationary seal that has a honeycomb surface. A serration cut honeycomb helps maintain close tolerances over a large temperature range. Fishmouth seals are circular, sheet metal, stationary, and interlocking. The use of these types of seals prevents excessive leakage of hot combustion gas from the primary airflow.

The cavity between the two seals is pressurized from aspirators (air ejectors) which are powered by eighth-stage air. The pressure in the pressurization cavity is always greater than the pressure inside the sump. For this reason air flowing from the pressurization cavity, across the oil seal, prevents oil from leaking across the seal.

The scavenge oil is drawn in from the sumps and transfer gearbox by the five scavenge elements of the pump, passes through the pump, through an outlet fitting on the enclosure base, and is returned to the Lube Oil Storage and Conditioning Assembly (LOSCA). LOSCA is a part of the lube system and contains all of the components of the scavenge system except the scavenge pump. Major components of the lubrication system are:

1. Lube oil storage and conditioning assembly
2. Lube and scavenge pump
3. Lube supply filter
4. Lube supply check valve
5. C and D sump supply check valve
6. Air/oil separator

Lube Oil Storage and Conditioning Assembly (LOSCA)

The LOSCA is located remotely from the gas turbine and includes an oil storage tank, a heat exchanger, a scavenge oil duplex filter, and a scavenge oil check valve. Two LOSCA are mounted in the engine room, one for each gas turbine. They are above the level of the lube and scavenge pump assuring a positive head for gravity feed to the pump. Synthetic oil flows through ship's piping between the gas turbine enclosure base and the LOSCA. The Main Reduction Gear (MRG) oil is used as the coolant for the LOSCA heat exchanger.

Oil Tank

The oil tank is an integral part of the lube storage and conditioning assembly. The early configuration of cast construction on the DD-963 contains six sight glasses (view ports) for visual detection of oil level in the tank. Start with the second glass from the bottom and space at 5-gallon intervals. On the later configuration of fabricated construction, three sight glasses are provided for low level, 19-gallon level, and full level positions. An oil level switch on the DD-963 monitors oil level from within the tank and transmits an electrical signal when the oil level is too high or too low. An oil level sensor on the DDG-51 monitors oil level from within the tank and transmits a continuous electrical signal for remote readout of lube oil level. The oil tank is considered full when oil is visible at the 24-gallon sight glass.

Instrumentation valves, a filter differential pressure transducer, a filter differential pressure gage, and an oil temperature sensor are mounted on the tank. A gravity filler cap is installed on the tank cover fill port. The fill port has a strainer to prevent foreign material from entering the tank. Baffles are located in the bottom of the tank to minimize oil sloshing. A deaerator is inside the tank at the scavenge inlet (which separates air from the scavenge oil). The oil tank may be drained by positioning a lever located in the assembly base.

Heat Exchanger

The heat exchanger (oil cooler) is a shell-tube assembly. The coolant is MRG lube oil (2190 TEP). The coolant passes from the MRG lube oil cooler through temperature control valves and through the inside of the tubes. The synthetic lube oil passes around the outside of the tubes. The end domes must be removed to gain direct access to the inside of the coolant tubes for cleaning.

Duplex Scavenge Filter

The duplex scavenge oil filter has provisions for manual selection of either element while the gas turbine is shut down or operating. The scavenge filter is located on and is part of the LOSCA.

Oil from the gas turbine enters the filter inlet port, flows through the selected element (outside to inside), and exits through the filter outlet port. A drain plug is at the bottom of each filter bowl so the oil may be drained from the element prior to its removal for cleaning. Filter selection is made by placing the selector lever in front of the element not being used. The spring-loaded locking pin must be raised from its locking slot before the lever can be moved. Engage the locking pin in the locking slot after the lever is positioned so the lever cannot move. A poppet relief valve is in the head of the duplex filter which opens at 24 psid to allow oil flow if filters should become clogged. The relief valve closes at 20 psid.

Scavenge Oil Check Valve

The check valve is between the scavenge filter and the heat exchanger. Its purpose is to prevent oil in the scavenge lines from draining back into the sumps and gearbox while the engine is shut down. The valve opens and flows at a rate of 20 gallons per minute (gpm) with a maximum differential pressure of 15 psid.

Lube and Scavenge Pump

The lube and scavenge pump is a six-element positive-displacement vane pump. One element is used for lube supply and five are used for lube scavenging. Scavenge oil enters the pump through five scavenge oil ports, passes through an inlet screen in each port, and enters the scavenge elements. The outputs of the five scavenge elements are connected inside the pump and discharge through a common scavenge discharge port. This scavenge discharge is routed to the duplex filter mounted on the lube storage and conditioning assembly oil tank.

The five scavenge elements of the lube and scavenge pump scavenge oil from the B, C, and D sumps and from two areas of the transfer gearbox. There are five resistance temperature detectors located in the scavenge oil system lines in the immediate vicinity of the pump. They sense oil temperatures for the A, B, C, and D sumps and for the transfer gearbox; they also provide a signal to the off-engine electronic controls.

Lube Supply Filter

The lube supply filter is a duplex type filter assembly with provision for manual selection of either element. It is the same type that is used for the scavenge oil filter. A relief valve in the filter opens to allow oil to bypass the filter if the filter becomes clogged. To make a filter selection, raise the spring-loaded locking pin, move the selector handle until it is in front of the element NOT being used, and release the locking pin. Make certain you engage the pin in the locking slot. A drain plug in the bottom of each filter bowl permits oil to drain from the element prior to removal for cleaning.

Lube Supply Check Valve

The lube supply check valve is on the downstream side of the supply filter. Its purpose is to prevent oil in the tank from draining into the sumps when the gas turbine is shut down. It opens and flows at the rate of 20 gpm with a maximum differential pressure of 15 psid.

C and D Sumps Check Valve

This check valve is in the lube supply line to the C and D sumps. Its purpose is to isolate the C and D sumps from the gas generator lube oil system when an external lube supply and scavenge system is used for the power turbine. Both the C and D sump oil supply lines and scavenge lines are equipped with fittings to facilitate the application of an external lube system for the power turbine. During normal engine operation, lube oil is supplied to the C and D sumps from the lube pump. The check valve opens at 2 psi of differential pressure.

Air-Oil Separators

The air-oil separator consists of a fabricated sheet metal impeller with a cast aluminum housing. To prevent excessive oil loss from venting oil vapor overboard, all sumps are vented to the air-oil separator. The sump air is vented to the exhaust duct after passing through the separator. Oil is collected on the inside of the impeller as the oil-laden sump air passes through the separator. Small holes in the segments of the impeller allow the collected oil to be discharged to the separator outer housing. Vanes on the housing wall are used to collect and direct the oil to the separator outlet where it is returned to the gearbox. To prevent oil and oil vapors from escaping past the end of the impeller, the separator has two labyrinth seals, with the cavity between the two seals pressurized with eighth-stage ejector air.

Gearbox and Seals

The transfer gearbox contains carbon oil seals that depend on a tight fit (rub) of the stationary member against the rotating member to prevent oil from crossing the seal. The stationary member is a carbon ring which is spring-loaded against a highly polished steel rotating mating ring. Oil is directed into the gearbox bearings by lube nozzles and is scavenged from the gearbox through a port located on the bottom aft face of the gearbox. Oil draining from the A sump (compressor front frame) flows into and is scavenged from the transfer gearbox.

The gearbox is vented through the radial drive shaft and drive shaft housing into the A sump cavity and out through the compressor front frame strut.

Gas Turbine Electronic Power Control System

The gas turbine electronic control system controls the amount of fuel that the propulsion turbine receives and thus the power output of the engine. The control system is the interface between the command signal from the ship's throttle and the Main Fuel Control (MFC) of the engine. The MFC can be thought of as a fuel oil valve where the throttle lever positions the valve controlling fuel flow to the engine.

The gas turbine electronic power control system or Engine Control Module (ECM) consists of a gas turbine-mounted electromechanical Power Lever Angle (PLA) actuator and a Free Standing Electronic Enclosure (FSEE). The FSEE generates the control signals to the PLA actuator, which provides the electromechanical interface with the MFC lever. The system performs the following functions:

1. Controls the PLA actuator that positions the power lever of the gas turbine MFC in response to a ship-provided electrical command signal.

2. Limits power turbine output torque by using various gas turbine parameters to calculate torque, comparing the calculated torque to a preset limit, and overriding the command signal to reduce the setting of the MFC power lever if the preset limit is exceeded.

3. Limits power turbine speed by comparing actual speed to a preset limit and overriding the command signal to reduce the setting of the MFC power lever if the preset limit is exceeded

4. Provides power turbine overspeed protection by monitoring two-speed signals from the power turbine and de-energizing the gas turbine fuel shutoff valves in the event of a power turbine overspeed or the loss of both speed signals

5. Limits power turbine acceleration by comparing actual acceleration rate to a preset limit and overriding the command signal to reduce the demand to the MFC power lever if the preset limit is exceeded.

PLA Actuator

The PLA rotary actuator is an electro-mechanical device and comprises a d.c. servomotor, a reducing gear, slide potentiometer, tachometer generator, mechanical linkage, and an electrical line filter. The PLA actuator is mounted on the fuel pump; it is connected to the fuel control power lever through a mechanical linkage. It is electrically connected to the FSEE. The purpose of the PLA actuator is to interface the PLA actuator electronics in the FSEE with the MFC of the gas turbine. Remember, the ultimate purpose of the PLA actuator is to move the MFT to control fuel flow to the engine.

Signals from the PLA actuator electronics, located in the FSEE, are converted by a servo-mechanism into mechanical action that positions the fuel-control power lever. Feedback of PLA and rate of change are sent to the FSEE. A positive mechanical rig feature allows locking of the PLA actuator output lever at a position of 113.5± 1°. This rig point is used in conjunction with a corresponding rig point on the fuel control. Its purpose is to establish mechanical synchronization between the PLA actuator and the fuel control.

Servomotor

The d.c. servomotor is driven by the PLA actuator drive signal from the PLA actuator electronics. The drive signal is developed by the electronics in the FSEE. It is amplified and connected through a relay referred to as the "fail-to-idle" relay. The motor's direction of rotation is determined by the polarity of the drive signal; its velocity is proportional to the drive signal amplitude. For an input range of -23 to +23 volts, the motor output shaft speed range is 0 to 900 rpm in either direction. The PLA actuator, after reducing by gear (ratio 55.64 to 1), has a speed range of 0 to 16 rpm (0 to 96°/second). It is also capable of running into mechanical stops at full voltage and velocity and remain stalled without damage to the motor.

Tachometer

The tachometer is a d.c. generator that is directly coupled to the motor shaft and outputs a d.c. voltage proportional to the motor speed. The polarity of the signal is dependent on the shaft's direction of rotation. The tachometer output range is 0 to 2.7 V d.c. nominal for a shaft input of 0 to 840 rpm.

Slide Potentiometer

The slide potentiometer provides a position feedback signal, which is proportional to the position of the PLA actuator shaft and MFC lever. It is a linear nonwire-wound variable resistance whose potentiometer slider position is controlled by the actuator output shaft. Two gears between the output shaft and the potentiometer increase the potentiometer's range of rotational movement by a factor of 2.27. The normal operating range of the actuator is approximately 100° and the potentiometer 227 °. The potentiometer reference voltages are supplied from electronics in the FSEE.

Free Standing Electronic Enclosure (FSEE)

The Free Standing Electronic Enclosure (FSEE) is located outside of the gas turbine enclosure. The enclosure is a metal cabinet that contains most of the electronics necessary for control of the propulsion turbines. The circuitry is divided into two identical sections—one for each turbine. The electronic circuitry is in the form of pull-out cards and a power supply rack; the cards may be referred to as either Printed Wiring Boards (PWB) or Printed Circuit Boards (PCB). Functionally, the electronics for either turbine can be divided into five subsystems: signal conditioning circuitry, torque computer, over speed switches, power supplies, and power lever angle actuator electronics. The FSEE also generates some signals (uplink) for display on the control consoles.

The FSEE on the DDG-51 class also contains the start/stop sequencer which provides for independent manual and automatic remote control of startup, operation, and shutdown of the gas turbine. The system also monitors various parameters to ensure safe gas-turbine operation. The same capabilities are provided for on the DD and CG classes of ships but from a separate system called the Engineering Control and Surveillance System (ECSS). Both the start/stop sequencer and the ECSS are covered separately. In the following discussion, we concern ourselves with the operation of the FSEE which, with the exception of acceleration limiting, are similar to all classes.

Signal Conditioning Electronics

The signal conditioning electronics is contained on one PWB—the E card. Five transducer signals and one internal signal are processed in these circuits. Four of these signals are provided as inputs to the torque computer. In addition, five signals are buffered and transmitted to external equipment. Two pressure signals, Compressor Inlet Pressure (P_{t2}) and power turbine inlet pressure (Pt5.4) are received in the form of 4 to 20 milliamp signals. They are converted to 0 to +5 volt signals for the torque computer. They are also processed into 0 to 10 volt signals for external use. The Compressor Inlet Temperature (T_2) is the only temperature signal processed by the FSEE. This signal comes from a platinum Resistance Temperature Detector (RTD). The RTD is a device which changes resistance with temperature. The electronics changes the RTD signal to a 0 to 5 volt and a 0 to 10 volt signal for the torque computer and uplink, respectively.

A dual tachometer system is mounted on the power turbine. This consists of a spur gear with 83 teeth which rotate past two sensors. The output of each sensor is a pulse train whose frequency is directly proportional to the speed of the turbine. The purpose for the two-speed signals is reliability. One is the normal signal, the other is a backup in case the first fails. The speed signal is converted to voltages proportional to speed for purposes of torque computer monitoring.

Torque Computer

The torque computer consists of 7 PWBs/PCBs whose only function is to calculate the torque output of the power turbine. Five inputs are necessary to the computer for the calculations. Four are 0 to 5 volt analog inputs from the signal conditioning circuitry $P_{,2}$, $P_{T5.4}$, T_{t2}, and Npf. The fifth is a discrete input to indicate whether bleed air is diverted from the gas generator (ON/OFF). This affects the efficiency of the engine, hence the torque output.

The torque computer is similar to many general purpose computers of this size. The computer calculates some internal values and compares them to internally stored tables. From the tables, selected values are taken to be used in further calculations to determine the final value of torque. The torque calculated is accurate to within 3000 ft-lb. When the turbine is at idle, the computer output is a torque value of approximately 5000 ft-lb. This is necessary within the electronics so it can operate accurately at other times.

The output of the torque computer is a 0- to 5-volt signal, which is the input to the PLA actuator electronics and a 0- to 10-volt signal for uplink and display. Torque range is 5000 to 50,000 ft-lb. The computer also calculates horsepower as a function of power turbine speed (N_{PT}) and torque for shipboard monitoring.

Overspeed Switch

Two PWB/PCB, the (D cards), of identical type are required to make up one power turbine overspeed switch. Speed signal channel A goes to one board, and speed signal channel B goes to the other board. Overspeed switch outputs consist of overspeed voltages (Limit 2), loss-of-speed signal voltages (Limit 1), and voltage for the fuel shutdown valves. Two test functions are provided.

1. Overspeed Test Function. Each board has a test generator with an output frequency above the highest power turbine speed to ensure a proper test. When you depress the test button, the overspeed indicator lights and the fuel shutdown valve de-energize.

2. Speed Limit Test Function. When you depress, the speed limit test pushbutton lowers the speed limit to 75 percent of the normal speed limit and permits testing the speed limit function without over speeding the power turbine.

If the power turbine speed signals should become disconnected or otherwise lost, the overspeed switch opens within 5 milliseconds. Both signals must be lost before the fuel shutdown valves actuate.

In order to start the gas turbine, a method of bypassing the no-signal shutoff is provided. With the throttle at a nominal 30 degrees or below, a signal is generated that causes the loss of speed-signal function to be bypassed. The throttle must remain below 30 degrees until the power turbine reaches 100 rpm. The signal loss bypass function is deactivated whenever the throttle is above a nominal 30 degrees. If fuel valve power is lost for longer than10 milliseconds, the fuel shutdown valves latch off to prevent reapplication of fuel to a hot engine. Loss of power is defined as any voltage level of 20 V d.c. or less. The overspeed switch receives its signal from the power turbine speed pickups, which are described in the signal conditioner electronics section.

Power Supplies

The power supply set is mounted on a pullout rack below the PWBs. It uses 28 V d.c. from PLOE and converts it to three regulated d.c. voltages for use in the FSEE: +5 V d.c, + 15 V d.c, and -15 V d.c. Other functions incorporated into the power supply are: the power amplifier for the PLA actuator signal, idle position signal relay used for uplink, and fail-to-idle relay used to bring the PLA actuator to idle by the system fail monitor.

PLA Actuator Electronics

The PLA actuator electronics consists of three PWB/PCB in the FSEE, the A, B, and C cards. The main function of this subsystem is to condition the command signal, compare it to the feedbacks from the actuator, and provide the signal to drive the PLA actuator to the commanded position. In conjunction with the main functions, this subsystem provides monitoring of certain system parameters for the protection of the gas turbine and power train.

The PLA actuator sends two signals back to the electronics. They are position feedback (potentiometer) and rate feedback (tachometer). In addition to the command signal, these two signals provide most of the information necessary for the electronics to place the MFC at the correct position.

The potentiometer receives its reference voltages from the PLA actuator electronics. The potentiometer slider takes a voltage proportionate to the actuator position from the potentiometer; this voltage signal is connected to the control circuit. The voltage signal is used to compare with the command signal and to generate an uplink signal representative of actuator position. Also, the position feedback is used within an idle position detection circuit; this circuit detects when the MFC is within two degrees of the normal idle position and generates an uplink signal to indicate when the MFC is at idle.

Rate feedback is developed by the tachometer attached to the motor shaft. The purpose of the rate feedback is to control the response of the PLA actuator during changes of MFC lever position. When there is a large difference between the commanded position and the actual position of the PLA actuator, the drive signal to the servomotor is large. A large drive signal causes rapid acceleration of the motor;

hence the motor could overshoot the desired PLA actuator position. Tachometer feedback tends to reduce this problem. If the feedback signal was not present, the desired position would be passed, forcing the system to backtrack by turning the motor in the other direction; it would eventually seesaw past the desired position a number of times before it settled at the correct position. The tachometer feedback anticipates the overshooting of the correct position and acts as a braking system for the motor. This is the same function that a compensating system in a hydraulic governor provides.

Protective Features

The protective functions of the PLA actuator electronics are referred to as torque limit control, speed limit control, acceleration limit control, PLA command rate limit control, and system fail monitor.

Torque Limit Control. The PLA actuator contains the circuitry that monitors for an overtorque condition. Torque is calculated for either split plant or full power on the DDG and CG classes of ships. If the torque signal received from the torque computer exceeds the limit, the limiting circuit goes into action to drive the MFC back, thereby reducing the torque output of the turbine(s).

Speed Limit Control. The speed limit control circuitry starts limiting when turbine speed reaches 3672 rpm. The purpose of the speed limit control is to keep the turbine speed below 3852 rpm. The circuit receives a speed signal from the signal conditioning card. This signal goes through an anticipation (to anticipate speed) amplifier; this amplifier detects the rate of increase of the speed (acceleration). This acceleration signal and the speed signal together are compared to a limit voltage. If that voltage is exceeded, the speed limiting circuit goes into action to limit the power turbine's speed or acceleration.

Acceleration Limit Control. A separate circuit also receives the power turbine speed signal where the rate of speed change (acceleration) is monitored. If the acceleration exceeds 332 rpm/second, this circuit lowers the PLA actuator drive signal to lower the power turbine acceleration.

System Fail Protection. The PLA actuator generates two signals when certain abnormal conditions are detected. One is an uplink signal to indicate a system failure; the other energizes the fail-to-idle relay to open the path of the PLA actuator's drive signal and inserts a fixed voltage to drive the MFC to idle.

If the command signal exceeds a maximum of + 12 V d.c. or falls below a minimum of + 0.3 V d.c, a command loss condition exists. Both the fail-to-idle relay energizing voltage and the system fail signal are generated. A malfunction is considered to exist if the unamplified drive signal to the rotary actuator exceeds +2.7 volts or below -2.7 volts for one second or longer. Again both the fail-to-idle relay energizing voltage and the system fail signal are generated.

If the MFC is at idle and an overtorque condition exists or if either of the 15-volt power supplies fail, a system fail signal is generated.

PLA ACTUATOR THEORY OF OPERATION

The command signal is developed by an operator at a throttle. When it reaches the PLA electronics, it first goes through a command rate limiting circuit that does not let the command signal into summation amplifier No. 1 until exceed a rate of change of 2.1 volts (22.5°)/second increasing or 9.0 volts (89°)/second decreasing. At summation amplifier No. 1, the command is compared to the position feedback signal. The difference is called the position error signal. The position error signal is passed through the controlled limiter attenuator. Here the signal is passed unchanged or is attenuated. Whether attenuation occurs or not depends on the limit discrete signal which is generated from within the limit loops when the torque or speed limit is exceeded. The attenuator attenuates the position error signal when the limit discrete signal is present. The attenuation of the position error signal lowers its effect on the PLA actuator.

Note: When the torque or speed limits are exceeded, the limit discrete and the analog limiting signals are simultaneously generated. The limit discrete signal lets the analog signal have more effect by diminishing the drive signal before the analog limiting signal is added to it at summation point No. 2.

The position error signal out of the attenuator along with the speed (PLA actuator rate) feed- back and the analog limiting signal from limit loops is added into summation point No. 2. There they are algebraically summed to generate the unamplified drive signal for the servomotor. The tachometer feedback signal tends to lower the drive signal whenever the PLA actuator moves fast. When the PLA actuator is moving slowly, the tachometer feedback is low and its effect is minimal on the drive signal. The analog limiting signal is generated in the limiting electronics when the torque, speed, acceleration, or PLA actuator rate limit is reached. This signal is proportional to the amount of PT torque, speed, acceleration, or PLA actuator rate present over their respective limit. It tends to drive the MFC in the direction necessary to remove the power turbine from the limit condition.

The corrected drive signal is sent through a power amplifier where it is amplified by a gain of 20. The signal then goes through the fail-to-idle relay to the servomotor. The fail-to-idle relay connects the drive signal to the PLA actuator through normally closed contacts; it may gate a fixed voltage, through normally open contacts, to drive the PLA actuator to idle if one of the conditions discussed previously exists.

The limiting functions of the PLA actuator electronics have a passive role during normal operation of the power turbine. Only when a limit is exceeded or, as in the case of speed limiting, seems like it is going to be exceeded, do the limiting circuits become active.

The inputs to the limiting circuits are power turbine speed and power-turbine torque. The speed signal comes from one of two power-turbine speed transducers through a signal conditioning circuit. The torque signal is calculated by the torque computer in the FSEE.

If more than one occurs simultaneously, the analog signals from the speed acceleration and torque limiting circuit are brought to the same point where the largest of the three gates to summation point No. 2 in the control loops. The analog signal from the rate limit can be applied to summation point No. 2 regardless of other inputs.

The operation of the control circuits in the PLA actuator electronics can be altered by the battle override function. This function can be used at the control consoles during testing and emergency operation of the ship. In the battle override mode, the fail-to-idle relay cannot be operated; therefore, the PLA cannot be forced to idle. The analog limit signal with which the limit loops can lower PLA actuator position is inhibited. Battle override can be activated for FSEE circuitry by activating the engine synchronizing switch located within FSEE.

FSEE Circuitry Tests. The electronics within the FSEE covers various internal tests to check if circuits are working properly. The tests are for the torque computer, the speed limit circuit, and the overspeed switch. It should be pointed out that these tests only test the FSEE circuitry. For a total system test, refer to the Planned Maintenance Subsystem (PMS).

Torque Computer Test. On this test several fixed parameter values are used by the torque computer to calculate a torque value. If the value exceeds the reference set point, an indicator light on the control console signals that the torque test passed.

Speed Limit Test. This test lowers the speed limit loop reference voltage by 25 percent. To test this loop using the turbines, the power turbine would only have to be run to 75 percent of real limit point.

Overspeed Test. There are four buttons— one for each channel in each turbine. A frequency generator in each PWB introduces a signal simulating a high-speed signal which causes the fuel valves to close and the overspeed light to illuminate.

INSTRUMENTATIONS

The GTM assembly is instrumented with sensors which provide for remote monitoring of the engine, module, and lube storage and conditioning assembly. Temperature, vibration and speed sensors give an electrical output signal directly, while pressure sensors use base-mounted transducers to convert a pressure level to a corresponding electric signal. The sensor information is transmitted to the controls, either directly or through the FSEE. The controls use the sensor information for GTM monitoring, alarming, and control sequencing.

Compressor Inlet Total Pressure (P_{t2}). Pressure is sensed by a total pressure probe mounted in the compressor front frame at the 12 o'clock position. Pressure is piped to a transducer mounted on the bottom of the enclosure base. The electrical output signal from the transducer is sent to the FSEE signal conditioner and torque computer electronics. Output signals from the FSEE are sent to the controls.

Power Turbine Inlet Total Pressure ($P_{t5.4}$). Pressure is sensed by five total pressure probes located circumferentially in the turbine mid frame. Pressure is piped to a transducer mounted on the bottom of the enclosure base. The electrical output signal from the transducer is sent to the FSEE signal conditioner and torque computer electronics. Output signal from the FSEE is sent to the controls.

Compressor Inlet Temperature (T_2). Temperature is sensed by a platinum Resistance Temperature Detector (RTD) penetrating the enclosure inlet barrier wall at the lower left corner. The signal from the RTD is sent to the FSEE signal conditioner and torque computer electronics. Output signal from the FSEE is sent to the controls.

Note: Only P_{t2}, $P_{5.4}$, and T_2 (which are torque computer inputs) are signal conditioned in the FSEE. Signals from the following instrumentation go to the controls where they are signal conditioned and used for monitoring, control, alarm, and shutdown as noted.

Power Turbine Speed (N_{PT}). Power turbine speed is sensed by two magnetic pickups in the turbine rear frame. The signal from the magnetic pickups is sent to the controls to the FSEE signal conditioner and torque computer electronics. The controls use the signal for PT speed meter displays and PT OVERSPEED alarm generation.

Gas Generator Speed (N_{GG}). Gas generator speed is sensed by a single magnetic pickup located on the top left side of the aft transfer (accessory) gearbox.

Gas Turbine Vibration. Vibration is sensed by two velocity pickups. One pickup is mounted on the compressor rear frame forward flange at the 12 o'clock position; one is mounted on the turbine rear frame forward flange at the 12 o'clock position.

Compressor Discharge Static Pressure (P_{53}). Pressure is sensed from a pressure tap on the P53 sensing line to the MFC and is piped to a base- mounted transducer.

Fuel Manifold Pressure. Pressure is sensed from a pressure tap on the fuel manifold downstream from the No. 2 fuel shutdown valve. The pressure is piped to a base-mounted transducer.

Fuel Pump Filter Differential Pressure. Pressure is sensed from two pressure taps in the fuel pump body at the filter inlet and discharge ports. Pressure is piped to a base-mounted transducer and a base-mounted gage.

Lube Supply Pressure (Pump Discharge). Pressure is sensed from a tap on the supply manifold downstream from the supply check valve. Pressure is piped to a base-mounted transducer.

Lube Supply Filter Differential Pressure. Pressure is sensed from taps in the filter head at the inlet and discharge ports. Pressure is piped to a base-mounted transducer and a base-mounted gage.

Lube Scavenge Filter Differential Pressure. Pressure is sensed from taps in the filter head and piped to a transducer. Both filter and transducer are mounted on the LOSCA.

Lube Scavenge Temperature. A, B, C, and D sumps and accessory gearbox temperatures are sensed by platinum RTDs installed in each scavenge line near the inlet ports of the pump.

Lube Scavenge Pressure. The pressure is sensed by transducer in the lube scavenge return line near the LOSCA.

Lube Cooler Outlet Temperature. Temperature is sensed by a platinum RTD installed in the cooler discharge line at the LOSCA storage-tank inlet port.

Fuel Inlet Temperature. Temperature is sensed by a platinum RTD installed in the fuel inlet line inside the gas turbine enclosure.

Power Turbine Inlet Gas Temperature ($T_{5.4}$). Temperature is sensed by 11 dual-element chromel-alumel thermocouples installed circumferentially in the turbine mid frame and electrically paralleled to produce a single output signal.

Enclosure Cooling Air Out Temperature. Temperature is sensed by a platinum RTD mounted on the enclosure ceiling on the centerline and just forward of the exit area.

Main Fuel Control (MFC) Power Lever Position. A 0.5- to 10-volt d.c. signal from the PLA actuator position feedback potentiometer is sent to the FSEE.

Summary

In this chapter and the previous one, you have seen how the LM2500 engine is constructed, and the function of its various parts. You have also followed the operations of its various systems including the flow of air into the combustion section, the mix of fuel and air, and how the fuel system and igniter system cause combustion. You have followed the operation of the lube oil system, and looked at a description of its purpose, and have seen how the engine is controlled electronically by the FSEE. We have briefly covered the control systems for the different classes of ships. We discuss the ECSS and start/stop sequencer control; the actual start, operation, and stopping of the engines in the chapter that follows. You can then see how the overall engine operation is controlled from the engine room or from a central control point aboard ship. By now you should be familiar enough with the engine and its operational systems to be able to follow instructions concerning basic maintenance procedures and understand the importance of the numerous parameters that control the engine.

CHAPTER 4

ALLISON 501-K17 GAS TURBINE GENERATOR

The Allison 501-K17 gas turbine engine is used aboard the CG-47 class of ship to provide electrical power. This electrical system consists of three gas turbine generator sets (GTGS), each of which contains a gas turbine engine, an alternating current generator, and associated equipment. The associated equipment includes the reduction gear, the control panel, and the enclosure assembly or module which houses the engine and reduction gear. Also associated with the gas turbine engine is a waste-heat boiler that uses the hot exhaust gases from the engine to provide steam for various services such as hot water and fuel and lube oil heating. Engineers are required to operate, maintain, and repair both the gas turbine engine and the waste heat boiler.

In this chapter you will study the enclosure assembly, the gas turbine engine, and the reduction gear. You will briefly study the generator; this includes its excitation and voltage control, the governor, the operation of the 501-K17 engine, and the seawater service system used for cooling. There have been various changes and modifications to the engine and controls on different ships such as DDG-51 class. The unit we discuss in this chapter is the model 104, which is the most common type found in the CG-47 class. However, further information on updates and modifications can be found in the class advisories and technical manual updates.

Allison 501-K17 Gas Turbine Generator Set

The GTGS consists of a gas turbine engine, reduction gear assembly, an ac generator, associated engine controls, and monitoring equipment. These components are mounted on a common base or contained within an acoustical enclosure.

Base Enclosure Assembly

This assembly consists of the mounting base for the engine, reduction gear, generator, acoustic enclosure with intake and exhaust system, and the cooling system that includes the water-wash and fire extinguishing system.

Base

The GTGS base is a steel structure attached to the ship's foundation through twelve 5000-pound capacity, shock-vibration isolating mounts. The base supports most of the GTGS system; exceptions are the generator exciter/voltage regulator unit (including the electronic governor) and a remotely mounted oil cooler for the gas turbine and the reduction gear lube oil systems.

Acoustic Enclosure

The gas turbine engine and the reduction gear assembly are housed in an acoustical enclosure. The enclosure reduces the noise level within the machinery space. Barrier walls within the enclosure separate the engine compartment from the reduction gear compartment; these walls form the inlet air plenum for the engine.

Blow-in and blow-out panels are provided to prevent damage to the enclosure in the event of high or low pressure within the enclosure. The panels are spring-loaded in the closed position. The blow-out panel is located in the enclosure roof panel on the left side and forward of the cooling air silencer. The blow-in panel is located on the lower aft corner of the enclosure's left side. The locations given are relative to an observer standing at the exhaust end looking towards the intake of the engine.

Included in the enclosure is an engine water-wash system. The water-wash system is used to clean the compressor section of the gas turbine. Two spray nozzles are mounted in the forward wall of the inlet plenum. They spray cleaning solution or fresh water into the engine inlet while the engine is being motored. With the exception of the spray nozzles and the solenoid-operated signal air valve, all components of the water-wash system are ship's systems.

Enclosure cooling air is provided via an air inlet silencer at the top of the module. Within this box are a louvered cooling air modulator, an axial fan, ducting to the enclosure, and a fire damper. Cooling air is extracted from the gas turbine intake duct and is directed through an enclosure baffle to the forward (compressor) end of the turbine. The air circulates around the engine and exits through a gap between the engine exhaust nozzle and the exhaust educator section where it mixes with the engine exhaust gases.

Intake and Exhaust System

The intake and exhaust system provides the flow path for combustion and cooling air to each gas turbine engine and each engine exhaust gas discharge. The inlet system contains inlet louvers, demisters, a blow-in door, and a silencer. The exhaust system contains a silencer and an infrared (IR) suppression system. There are three engines aboard ship with an engine intake and exhaust system for each engine. The exhaust gas from each engine is routed through the waste-heat boiler before entering the exhaust stack.

Intake Duct. The intake duct is a rectangular structure. Air enters the duct through louvers mounted in the side of the stack, flows through the mesh-pad demister, through the silencer, into the module inlet plenum, and into the engine inlet. The intake air inlet for the number 3 GTGS is located on the 01 level, starboard side, aft of the missile launcher area. Air enters a vertical bellmouth and flows downward into the number 3 generator inlet plenum. This plenum serves as a green water trap and allows any large quantities of water to drain through slots in the deck combing. The air then flows through demisters into the number 3 generator intake room. The bulkhead between these two compartments contains the blow-in door. Combustion and cooling air flow through separate ducts from the intake room to the module.

The intake duct inlets for the numbers 1 and 2 engines have louvers similar to the main engine inlet louvers. Like the main engine louvers, they are designed and arranged to shed sea spray but are not heated. Because of the vertical flow inlet configuration, the number 3 engine duct inlet has no louvers.

The demisters are the mesh-pad type, similar to those in the main engine inlet. They are arranged vertically in a room behind the louvers. Moisture separated from the air collects in scuppers under the demisters and is drained overboard.

A single blow-in door is located in each inlet below the demisters. The purpose of the doors is to bypass the demisters if they become clogged to permit sufficient combustion and cooling air flow to the engine for normal operation. A controller provides for manual or automatic operation by a selector switch on the controller door. When operated in manual, a push button on the controller door energizes a solenoid and releases the blow-in door. When in automatic, the solenoid energizes by action of a pressure switch set to operate at approximately 6-8 inches of water differential pressure. Indication and alarm of DUCT PRESS LO are given at Propulsion Local Operating Equipment (PLOE) and the Propulsion and Auxiliary Machinery Control Equipment (PAMCE). Once open, the door must be closed manually.

The vane-type silencers consist of sound-deadening material encased in a perforated, stainless steel sheet. They are mounted vertically in the duct between the demisters and the cooling air duct.

Exhaust Duct. The exhaust duct is a round, insulated, stainless steel structure. Each duct contains a silencer, an eductor, and an

IR suppression system. The eductor provides an outlet for module cooling air and aids in cooling by sucking the module air into the exhaust stream.

The exhaust ducts from the numbers 1 and 2 engines run parallel to the main engine ducts in the exhaust stacks. The duct from the number 3 engine traverses the ship and discharges from the port side away from the king post and weapons platform on the 01 level aft.

The silencer consists of sound-deadening material encased in a perforated stainless steel sheet cylinder which is suspended in the center of the exhaust duct. This unit, together with the duct wall insulation, provides the required sound reduction to meet the airborne noise requirements.
Anti-Icing

The anti-icing system is similar to that in the main engine inlet ducts. The hot bleed air from the engine discharges into the inlet duct where it mixes with the inlet air and raises the temperature above the freezing point. Bleed air flow is regulated as a function of upstream temperature versus a fixed temperature to maintain an inlet temperature of approximately 38 °F when anti- icing is selected. This temperature is sufficient to prevent the formation of ice and to melt any ice, sleet, or snow entrained in the air.

Fire Detection and Extinguishing System

Each GTGS module is protected by independent primary and secondary systems; each contains individual banks of CO_2 cylinders, discharge piping systems, and controls. The primary system provides rapid flooding of the module; when activated, it releases the entire contents of the primary bank at a rate of 200 pounds per minute. Release may be initiated automatically by the fire detectors inside the module or manually. The secondary system sustains an inert atmosphere inside the module at a rate of 67 pounds per minute. The secondary system is normally released only after the primary system has been discharged and must be released manually at the bank. Once released, the entire content of the bank is discharged into the module.

The primary system consists of two 50-pound CO2 cylinders, two pressure switches, and two high-volume, low-velocity nozzles. The CO2 cylinders are mounted in racks adjacent to the module. The pressure switches are located in the piping system, one outside and the other inside the enclosure. The nozzles are mounted on the air intake assembly. The GTGS fire detection system consists of two flame detectors mounted on the enclosure wall. When flame is sensed by the flame detectors, a fire signal is transmitted to the cooling fan controller where the fire alarm and shutdown relays are located. After a 5-second time delay, the primary CO2 bank is released. Simultaneously, a stop command is transmitted to the GTGS start/stop logic, the module cooling air fire damper closes, the cooling fan turns off, the fuel valve closes shutting down the engine, 5th- and 10th-stage bleed air valves close, and are held closed by ship's service air, fire alarms activate at the Damage Control Console (DCC) and Electric Plant Control Enclosure (EPCE) and a summary alarm activates at the switchboard. The GTGS start/stop logic ensures that 14th-stage bleed air is shut off when gas turbine speed is below 12,780 rpm. Two interlock switches are operated by the enclosure access doors; these switches pre- vent actuation of the CO2 release solenoid when the doors are open.

The primary bank can also be activated manually at the bank or remotely from the pull box. When the two pressure switches are operated by CO2 pressure in the header, a CO2 release alarm activates locally and at the DCC. The summary alarm at the switchboard is also activated. Once released, CO2 discharge cannot be stopped. Again the primary discharge is at the rate of 200 pounds per minute.

The secondary system consists of three 50-pound CO2 cylinders and two high-volume, low-velocity nozzles connected together by a common piping system. The secondary bank must be released manually at the bank. The secondary system is not equipped with monitors or alarms. Once released, CO2 discharge cannot be stopped. Again the secondary discharge is at the rate of 67 pounds per minute.

GAS TURBINE ENGINE ASSEMBLY

The Allison Model 501-K17 is a single-shaft, axial-flow gas turbine that has a 14-stage axial flow compressor, a cannular combustor, and a 4-stage axial-flow turbine directly coupled to the compressor. The gas turbine drives the generator through a reduction gear. The reduction gear is mounted in front of the gas turbine and is coupled to the compressor front shaft by a power Take-Off (PTO) shaft. The gas turbine and reduction gear are mounted in a common shock-mounted, sound-attenuating enclosure. The gas turbine is mounted on a suspension system at the approximate center of gravity. It is adjusted so that minimum stress is placed on the bolted flanges of the PTO housing. This allows freedom of movement in all planes and maintains engine reduction gear alignment when deflection occurs due to shock, thermal growth changes, and so forth. Rotational direction of the engine is counterclockwise when viewed from the exhaust end.

Air Intake

The air intake section consists of a one-piece cast-aluminum housing which forms the airflow path to the compressor. The air inlet housing consists of an outer case, a center hub, and eight struts connecting the hub to the outer case. Mounted on the outer shell at the 12 o'clock position is an engine breather and at the 6 o'clock position is the accessory gear box. The struts between the outer and inner shells are drilled for oil and air flow through them. The bottom strut contains passages for pressure oil, scavenge oil drain, and the accessory gearbox radial drive shaft. This shaft transfers power from the compressor rotor to the accessory gearbox. The inner shell houses the compressor extension shaft housing, the compressor extension shaft, and the compressor front bearing.

The compressor extension shaft housing is secured in the base of the air inlet housing. The two housings form two annular passages sealed by three O-rings. The compressor extension shaft is supported at the forward end by the compressor extension shaft bearing which is a ball-type bearing. The rear end of the shaft splines into the 1st stage wheel hub. The PTO shaft is splined into the forward end of the extension shaft. A bevel gear on the extension shaft drives a side gear with which the accessory gearbox radial drive shaft is splined.

Lubrication of the bearings for this area is accomplished through the use of jet spray nozzles. These nozzles receive the oil from the rear annular passage formed by the compressor extension shaft housing and the air inlet housing inner shell. Oil from a nozzle on the aft face of the extension shaft housing sprays on the compressor front bearing. A jet, drilled into the compressor front bearing oil passage, sprays lubricant on the extension shaft bearings. Another nozzle on the front of the housing sprays oil on the PTO mid-bearing.

Compressor Section

The compressor section consists of a compressor stator. A compressor rotor and a diffuser are in addition to the air inlet housing described in the proceeding paragraph. The compressor case is made up of four sections bolted together along horizontal split lines. On the inside diameter of the casing, machined channels provide for the mounting of the compressor vane assemblies of stages 1 through 13. The compressor rotor assembly is encased by the compressor casing assembly. The rotor is supported at the front end by a roller bearing, and, on the aft end, it is supported and axially positioned by a ball-type bearing. A compressor tie bolt extends through the hubs of the rotor wheels and threads into the hub of the 14th-stage wheel to hold the rotor together. The compressor casing assembly contains a series of holes at the 5th and 10th stage. Manifolds are mounted on the outside of the casing to collect bleed air.

The diffuser is a fabricated steel assembly which is designed to offer a minimum amount of restriction to the flow of compressed air into the combustion section. The 14th-stage compressor and outlet vane assembly receives air from the rotor and directs it at the proper angle into the diffuser. The diffuser is bolted to the compressor casing at the forward end and to the combustion chamber outer casing. It is made up of an outer shell connected to an inner cone by six hollow struts. The aft side of the struts has holes to extract some of the air flow and direct it into an annular passage which feed four diffuser radial struts. Three of these radial struts are connected to a manifold to provide 14th-stage bleed air.

Located above and below the two horizontal are four holes in the outer shell. The hole above the right strut supplies a pressure signal for fuel control during start up Compressor Discharge Pressure (CDP). The hole below the left strut supplies the operating medium for the 5th- and 10th-stage valves. The remaining two are not used.

There are six fuel nozzle mounts and openings located between the radial struts of the diffuser. The fuel nozzles, secured to these mounts, support the forward end of the combustion liners. Pressure oil for the compressor rear bearing and turbine front bearing enters a tube just forward of the upper right strut.

Combustion Section

The combustion section consists of six individual combustion chambers (burner cams) equally spaced in an annulus formed by a one-piece outer casing and a two-piece inner casing. The outer casing is a fabricated assembly which bolts to the diffuser and turbine inlet casing. The casing provides mounting for six equally spaced liner supports. The liner supports extend into the combustion liners and axially position and retain the combustion liners within the outer combustion casing. The liner supports in combustion liners two and five house the two spark igniters. Two burner drain valves are used to drain unburned fuel after shutdown or a false start.

Six identical combustion liners are housed between the outer and inner casing. In addition to being axially positioned by the liner supports, the combustion liners are positioned radially by the 1st stage assemblies in the turbine inlet casing.

Liners are designed to control the length and position of the flame within the liner and to provide a rapid fuel-air mixing area. This is accomplished through the use of swirl vanes at the forward end of the liner around the fuel nozzles for fuel air mixing. Further down the side of the liner are primary air holes which are also used in the combustion process. Around the center portion of the wall are the secondary air holes and reverse flow baffles. This air is used to center the flame and control its length and cool the liners. Reverse flow baffles redirect the secondary air toward the forward end of the liner. Because of these baffles these cams are classified as smokeless liners. These liners end by crossover tubes which are used to equalize internal pressures and to provide a path for flame propagation during start.

The inner combustion casing is a fabricated steel assembly which is bolted to the turbine inlet casing and slip fits into the diffuser inner combustion casing sleeve. An alignment bellows, located near the aft end, allows for slight misalignment between the turbine inlet casing and diffuser without damage to the inner combustion chamber.

The inner combustion liner is a fabricated steel assembly which is bolted to the inner combustion casing at the forward end and to the turbine inlet casing at the rear. Near the aft end of the liner is an expansion joint to allow for thermal expansion of the inner casing. Inside the casing are two oil tubes. The upper tube is pressure oil to the turbine front bearing. The lower tube is scavenge oil from the turbine front bearing sump.

The turbine coupling shaft passes through the interior of the inner combustion casing liner. It is used to transmit turbine rotor torque to the compressor. The front of the shaft splines into the compressor coupling and into the 14th-stage compressor hub. The aft end of the shaft splines into the turbine shaft coupling assembly.

Power Turbine Section

The turbine section consists of the turbine inlet casing, the turbine vane casing, and the rear bearing support. The turbine assembly receives hot gases from the combustion section and, by means of the 4th-stage turbine, converts the energy of the expanding gases into torque for driving the compressor, accessories, and generator.

The turbine inlet casing is a fabricated member made up of an inner and outer shell connected by six hollow radial struts. The outer shell bolts to the combustion outer casing at the forward end and to the turbine vane casing at the aft end. Eighteen thermocouple units are mounted on the outer shell, with their probes extending into the exhaust gas path of the combustion liners.

The 1st stage vane assembly is mounted inside the outer shell and supported by the vane support. The vanes are hollow and are cooled by secondary air. The aft end of the combustion liners section fits over the inner and outer rims of the 1st stage vane segment assemblies to support and position the liners. The inner shell houses the bearing sup- port and the bearing cage. The inner combustion casing bolts to the inner shell, while the inner combustion casing liner is attached to the bearing shell. The turbine vane casing is secured to the inlet casing outer shell forward and the rear bearing support aft.

The turbine rear bearing support is a fabricated structural member consisting of an outer shell and an inner shell connected by seven tangential struts. It is secured to the vane casing and houses the turbine rear bearing. The outer shell rear section forms the exhaust nozzle of the engine. An inner front exhaust cone and an inner rear exhaust cone serve to streamline the exhaust gases as they exit the engine. Retained in the inner shell is the rear bearing cage, the oil seal assembly, and an oil spray nozzle. The turbine rear bearing oil tube penetrates the outer shell just forward of the lower left strut and follows the strut to thread into the oil seal assembly to lubricate the turbine rear bearing. Inside the inner rear exhaust cone and attached to the inner shell is the turbine rear scavenge pump.

The turbine rotor assembly consists of the four turbine wheels with blades, three turbine wheel spacers, curvic seal rings, the turbine vent tube, and eight turbine clamp bolts.

The 4th-stage turbine rotor assembly is radially supported by two roller bearings and is positioned and retained axially by the turbine to compressor tie bolt. The four turbine wheels are held together by eight clamp bolts. The turbine wheels have machined shoulders near the inner diameter which have curvic splines cut in them. These splines transmit the torque produced in the turbine.

The turbine rear scavenge pump is mounted in the pump support inside the inner rear exhaust cone. The scavenge pump drive shaft coupling is splined into the aft end of the turbine to compressor tie bolt. The oil pump drive shaft splines into the aft end of the shaft coupling and has a bevel gear near its aft end which drives the turbine rear scavenge pump.

Accessory Drive

The accessory drive assembly (accessory gearbox) provides mounting pads on the rear face for the fuel pump, governor actuator, and external scavenge oil pump. Pads on the front face are for the speed-sensitive valve, main oil pump, and oil filter. The accessory drive assembly is driven by the compressor rotor extension shaft. Bevel gears, located in the inlet housing, drive the radial shaft. The radial shaft is located in the bottommost strut of the inlet housing and is connected to the accessory gearbox.

Ignition System

The ignition system consists of an ignition exciter, high-tension leads, and two spark igniters. The system operates on a nominal positive 28 volts d.c. However, satisfactory operation can be obtained over a range of positive 14 to positive 28 volts dc. Power is supplied to the system through an electronic speed switch actuated relay, which energizes the system at 2200 rpm and deenergizes at 8400 rpm during the starting cycle.

Ignition Exciter. The ignition exciter is a hermetically sealed (airtight) unit mounted on the right side of the compressor. It is a high-voltage, capacitor-discharge exciter capable of firing two spark igniters at the same time.

Spark Igniters. The two spark igniters are mounted in the outer combustion case, one in the number 2 can and one in the number 5 can. The igniters receive the electrical output from the ignition exciter and discharge this electrical energy during starting to ignite the fuel-air mixture in the combustion cans. Two high-voltage leads connect the spark igniters to the ignition exciter.

Bleed Air System

The bleed air system is two independent systems; 14th-stage system and 5th- and 10th-stage systems. The purpose of the 5th- and 10th-stage bleed air systems is to unload the compressor to reduce the possibility of compressor surge (stall) during the starting cycle. The purpose of the 14th-stage bleed air system is to extract air from the compressor for the ship's bleed air system. Airflow up to 2.37 pounds per second at 55 to 60 psig may be extracted; this is approximately 10 percent of compressor airflow.

Fourteenth-Stage Bleed Air System. The 14th-stage compressor discharge air is extracted from ports on the compressor diffuser, manifold and piped to a bleed air control valve, and into the ship's bleed air system. The control valve, upon a signal from the gas turbine generator control panel, prevents the system from being overloaded during combined operation of bleed air and generator loading. If the turbine inlet temperature (TIT) reaches 1870°F, the bleed air control valve closes to a point that maintains the TIT in the range of 1850 °F to 1870°F. The manually operated 14th-stage bleed switch is located on the gas turbine generator control panel. It allows the operator to enable the bleed air control circuit. When this switch is in the ON position, the bleed valve opens at 12,780 rpm and is fully automatic with respect to TIT.

Fifth and Tenth-Stage Bleed Air System. This system consists of eight pneumatically operated bleed air valves, a speed- sensitive valve, a filter, and a two-way solenoid valve. Four bleed air valves are mounted on both the 5th- and the 10th-stage bleed manifolds. These valves are piston-type valves, with 5th- and 10th- stage air pressure on the inboard side of the valve and either atmospheric pressure or 14th-stage air pressure on the outboard side. The speed-sensitive valve is engine driven and is mounted on the forward side of the accessory gearbox. The valve has three ports. One port is piped to 14th-stage air; one port is piped to the outboard side of the 5th- and 10th-stage bleed air valves; the third port is vented to atmosphere. During operation at engine speeds below approximately 12,700 rpm, a pilot valve in the speed-sensitive valve is positioned in a way that the 14th-stage air is blocked and the outboard side of the bleed air valves are vented to atmosphere. Since the 5th and 10th stage air pressures are greater than atmospheric pressure, the valves open and vent air from the compressor. At engine speeds above approximately 12,700 rpm, the pilot valve is in a way that the vent port is closed and 14th-stage air is ported to the outboard side of the bleed air valves. Since the 14th-stage air pressure is greater than 5th- and 10th-stage air pressure, the valves close and bleeds air is stopped. The filter is located in the 14th-stage air line to the speed-sensitive valve to prevent contaminants in the air from clogging the valve. The solenoid valve is located in the line between the speed-sensitive valve and the bleed air valves. It uses ships service air to hold the 5th- and 10th-stage bleed air valves closed during a fire or while water-washing the engine.

ENGINE FUEL SYSTEM

The fuel system meters and distributes fuel to the engine to maintain a constant rotor speed under varying generator load conditions. Components of the fuel system are both engine mounted and off-engine mounted

The engine-mounted components include a dual-element fuel pump, low-pressure fuel filter, high-pressure fuel filter, pressure relief valve, liquid fuel valve, electrohydraulic governor actuator, fuel enrichment valve, fuel shutoff valve, manifold drain valve, fuel enrichment pressure switch, fuel nozzles, and burner drain valves.

Off-engine mounted components include a temperature biased Compressor Inlet Temperature/Compressor Discharge Pressure (CIT/CDP) sensor and a start temperature limit control valve.

In operation, fuel from a gravity feed tank enters the enclosure and flows into the inlet of the fuel pump, passes through the pump boost element, through the low-pressure filter, and into the high-pressure elements. From the pump's high-pressure elements, the fuel passes through the high-pressure filter and into the liquid fuel valve. Metered fuel from the liquid fuel valve passes through the fuel shutoff valve, and then through the fuel manifold to the fuel nozzles where it is discharged into the combustion chambers. Since the fuel pump delivers more fuel than is required, the liquid fuel valve bypasses the excess fuel back to the inlet side of the pump's high-pressure elements. Additionally, some of the fuel from the pump's high-pressure elements is piped around the liquid fuel valve, through the solenoid-operated fuel enrichment valve, and into the metered fuel line between the liquid fuel valve and the fuel shutoff valve. The fuel enrichment valve and pressure switch has been deleted as an operational function of the engine due to the possibility that a leaky valve could cause a start overtemp during start or cause hot section damage during operation. The fuel enrichment feature is a carryover from the aircraft design and serves no useful purpose in the engines present configuration.

Fuel Pump. The fuel pump is an engine- driven, dual-element pump mounted on the aft right side of the accessory gearbox. The boost element consists of an impeller centrifugal pump and bypass valve. The high-pressure element consists of a dual-element (primary and secondary) gear pump. In operation, fuel enters the pump, passes through the low-pressure filter, and returns to the high-pressure elements through passages in the high-pressure filter assembly. The bypass valve opens only in the event of boost pump failure. Fuel normally flows in series through the secondary and primary elements of the gear pump. However, the two elements are placed in parallel from approximately 2200 rpm to 8400 rpm by a solenoid-operated paralleling valve, located internal to the fuel pump assembly. From the high-pressure element of the pump, fuel flows to the high-pressure filter.

Low-Pressure Filter

The low-pressure filter is a paper cartridge located in the fuel line between the boost pump outlet and the high-pressure element inlet. Relief valves are incorporated in the filter head which bypasses the fuel in the event the filter becomes clogged. Low-pressure filter inlet and outlet pressures are indicated on the engine gauge panel.

High-Pressure Filter

The high-pressure filter assembly is mounted on the bottom of the fuel pump and consists of a filter, a bypass valve, two check valves, and a solenoid-operated paralleling valve. The filter is a 33-micron disc, removable for servicing. The bypass valve opens to permit continuous flow if the filter becomes clogged. In the event one high-pressure gear element fails, the check valves permit engine operation from the other element.

Pressure Relief Valve

The pressure relief valve is closed during normal engine operation. If the pump discharge pressure reaches 500 ±10 psig above bypass line pressure, the relief valve opens to permit excess fuel to return to the pump.

Liquid Fuel Valve

The liquid fuel valve is mounted on the left side of the engine. It is mechanically and hydraulically connected to the electrohydraulic governor actuator. The hydraulic connection is through the CIT/CDP sensor and the start temperature limit control valve. It consists of a metering valve, acceleration limiter, and bypass valve. Its purpose is to meter the required fuel for all engine operating conditions. The electro- hydraulic governor actuator linkage and the acceleration limiter (internal part of the fuel valve) control the metering valve position, hence, the fuel flow. The acceleration limiter schedules fuel flow during starting as a function of compressor discharge pressure and compressor inlet temperature. During starting and rapid acceleration the limiter overrides the governor input, thus preventing compressor surge (stall) and excessive TIT. The limiter linkage (internal) is actuated by servo oil pressure from the electrohydraulic actuator, which is regulated by the CIT/CDP sensor.

Fuel flow is metered accurately by maintaining a constant pressure drop across the metering valve. To maintain a constant pressure drop across the metering valve, excess fuel from the pump is returned to the pump by the bypass valve.

The fuel enrichment valve and pressure switch has been deleted as an operational function of the engine due to the possibility that a leaky valve could cause a start overtemp during start or cause hot section damage during operation. The fuel enrichment feature is a carryover from the aircraft design and serves no useful purpose in the engine's present configuration.

Fuel Shutoff Valve

The fuel shutoff valve is a normally closed, solenoid-operated valve located in the line between the liquid fuel valve and the fuel manifold. All fuel to the fuel nozzles must pass through this valve. During the starting cycle, the valve is opened (energized) by the electronic speed switch circuit at approximately 2200 rpm. The valve is closed by the control circuits to shut down the engine.

Fuel Manifold Drain Valve

The fuel manifold drain valves are spring-loaded, solenoid-operated valves located at the bottom of the manifold. Their purpose is to drain the fuel from the manifold to the waste oil drain tank during coastdown. The valve is open (energized) only during the 2-minute period determined by the coastdown timer.

Electro-Hydraulic Governor Actuator

The electrohydraulic governor actuator is engine driven; it is mounted on the left side of the accessory gearbox. Its output shaft is mechanically linked to the liquid fuel valve. It receives signals from the Electric Governor (EG) control box and positions the liquid fuel valve, which in turn meters fuel to the engine. The governor actuator incorporates normal control by the electronic governor system, and back-up control by a centrifugal governor, each independently capable of positioning the output shaft.

An integral oil pump provides servo oil pressure for governor operation as well as other functions. Engine lube oil pressure from the accessory gearbox is supplied to the actuator pump through an external line. During normal operation, an output signal from the EG control box produces a force on an armature magnet attached to a pilot valve plunger, moving the plunger up or down. The pilot valve plunger directs servo oil pressure to change the position of the output shaft. If the electrical signal to the governor actuator is interrupted or attempts to overspeed the engine, the pilot valve plunger and terminal shaft positions itself towards the maximum fuel flow position. When the engine speed exceeds the preset limit, the centrifugal governor assumes control of the engine. Flyweights, opposed by speeder spring force, position the pilot valve plunger as a function of engine speed. The pilot valve plunger directs servo oil pressure to position the output shaft connected to the liquid fuel metering valve. The centrifugal governor is set to regulate engine speed at approximately 480 to 580 engine rpm above the normal electric governor operating speed, at full load, having been factory adjusted between 14,300 to 14,400 rpm.

Compressor Inlet Temperature/Compressor Discharge Pressure (CIT/CDP) Sensor

The CIT/CDP sensor senses both CIT and CDP and regulates servo oil from the electrohydraulic governor to the acceleration limiter in the liquid fuel valve as a function of CIT and CDP. The acceleration limiter in turn schedules fuel flow as a function of CIT and CDP. During the start cycle above 2200 rpm and during rapid accelerations, the acceleration limiter overrides the input from the electrohydraulic governor, limiting the maximum fuel flow, and thereby preventing compressor stall and/or excessive TIT. Below 2200 rpm, the regulated oil pressure from the CIT/CDP sensor is blocked by the start temperature limit control valve to assure the turbine starts on the minimum fuel flow at light-off. The CIT/CDP sensor is mounted on the inlet air plenum and the temperature sensing element protrudes into the inlet airstream.

Fuel Manifold and Nozzles

The fuel manifolds consist of sections of steel-braided hose, connecting the six fuel nozzles together and connecting them to drain valves at the bottom of the engine. Fuel output from the fuel flow divider valve is also connected to the primary and secondary manifolds. The fuel nozzles are air blast type. They consist of a body holder, primary port, secondary port, filter screen, filter screen spring, check valve assembly, primary spray tip, and an air blast shroud. When fuel manifold pressure is approximately 150 psig or less, the fuel flow is through the primary spray tip. This creates a spray pattern for starting and for stable combustion at low engine speeds. When fuel manifold pressure exceeds approximately 150 psig, the flow divider valve redirects fuel into the secondary manifold, secondary ports and orifices, diagonally outward from the air blast shroud swirlers. The fuel mixes with compressor discharge air flowing from the swirlers to form a finely atomized fuel spray pattern.

Burner Drain Valves

Two burner drain valves, located in the bottom of the combustion chamber outer casing, drain residual fuel from the combustor at shutdown or after a false start. The valves are lightly spring-loaded open; they close by combustor internal air pressure during engine operation. These valves open when combustion pressure drops below 1 to 5 psig and close above 1 to 5 psig on increasing pressure.

Start Limit Control Valve

The start limit control valve is a normally open, three-way solenoid-operated valve located in the regulated servo oil supply line between the CIT/CDP sensor and the liquid fuel valve. During the start cycle below 2200 rpm, the valve energizes, blocking the regulated oil supply and porting the oil from the acceleration limiter (part of fuel valve) to drain. This causes the fuel valve to remain against the minimum fuel flow stop until the engine reaches 2200 rpm. Between 2200 rpm and 12,780 rpm, the valve is normally deenergized (open). However, if TIT exceeds 1500°F, the valve is intermittently energized/deenergized until temperature drops below 1500°F. Above 12,780 rpm, the valve is electrically locked out of the system (deenergized).

A modification is being made to the fuel system to improve starting, acceleration and load transient response. This system is called the closed loop fuel system. It eliminates the CIT/CDP sensor and the start temp limit control valve. During start and acceleration the fuel scheduling is controlled by the governor actuator. The CIT/CDP sensor has been replaced by an RTD type of CIT sensor which inputs the electronic governor control box. The control circuits have been replaced by the Model 9900-320 governor control system. This system also receives electrical signals from the Allison Speed temp box for TIT, and a dual-type speed pickup mounted on the PTO shaft for engine speed. The liquid fuel valve has been modified to receive direct CDP pressure, and also incorporates a feedback circuit to the electronic control unit. This closed loop system enables the fuel to be scheduled by the governor actuator during all phases of generator operation. This system improves fuel scheduling by eliminating problems that have occurred due to oil leaks, dirty orifices, and malfunctioning valves. If your ship has received this modification, consult your technical manuals for further operation.

ENGINE LUBRICATION SYSTEM

The engine and reduction gear lube systems share a common supply tank, filter, and cooler. The supply tank is the reduction gear sump; the filter is base-mounted inside the enclosure. The oil cooler is mounted remotely from the module. Synthetic oil, MIL-L-23699, is used in this system.

The engine lube system is a low-pressure, dry-sump system incorporating a combination lube and scavenge pump, an external scavenge pump, a turbine scavenge pump, a pressure regulating valve, an oil filter and check valve, a filter bypass valve, and a scavenge pressure relief valve.

In operation, oil from the reduction gear sump (supply tank) is picked up by the reduction gear supply pump, flows through the supply filter, then through the oil cooler. Oil from the cooler is regulated to 25 psig to lubricate the reducing gear and supply oil to a 15 psi regulator. This oil will go to either the PTO shaft coupling or supply the engine main lube oil pump with a minimum of 15 psig oil. From the engine lube oil pump, where pressure is increased to above 60 psig, the oil flows through the L/O filter to the pressure regulator, where pressure is regulated to 55 ±5 psig, to drilled passages and external piping to the engine bearings.

Scavenge oil is picked up by the scavenge element of the main lube and scavenge pump, the external scavenge pump, and the turbine scavenge pump. Oil from the turbine scavenge pump flows through drilled passages and internal lines to the accessory gearbox, where it is picked up by the scavenge element of the main pump. Flow from the external scavenge pump joins the flow from the main scavenge pump through external lines and is returned to the reduction gear sump. The magnetic drain plugs (not shown) are provided on the bottom of the accessory gearbox and the discharge of the main scavenge pump to collect any steel particles in the oil.

Main Pressure and Scavenge Oil Pump

The main pressure and scavenge oil pump assembly is mounted on the front of the accessory gearbox. It consists of a pressure regulating valve and two gear pumps (one each for the supply and scavenge systems). Oil is pumped by the pressure (supply) element of the pump to the compressor extension shaft bearing, the power takeoff shaft mid bearing, the accessory gearbox, the engine four main bearings, and the electro-hydraulic governor actuator. The main shaft splines are lubricated by the oil returned by the rear turbine scavenge pump. The scavenge element picks up scavenge oil in the accessory gearbox, which has gravity drained from the compressor extension shaft bearing, and the compressor front bearing; this oil and the oil from other scavenge pumps then returns to the reduction gear sump. An indicating magnetic plug is located in the scavenge side of the pump.

Oil Filter. An oil filter is mounted on the front of the accessory gearbox. It has a pleated element and incorporates a Teflon-seated, poppet check valve to prevent oil from draining into the engine when the engine is shut down. A bypass valve is located in the accessory gearbox front cover; this valve opens at a specific pressure differential to bypass the filter in the event it becomes clogged.

External Scavenge Pump. The external scavenge pump is a three-gear, dual-element pump mounted on the aft side of the accessory gearbox. It scavenges the oil from the compressor rear bearing sump and from the turbine forward bearing sump. The oil from the pump is combined with the scavenge oil from the main scavenge pump and returns to the reduction gear sump.

Turbine Scavenge Pump. The turbine scavenge pump is mounted in the gear turbine bearing support assembly; it is driven by the turbine-to-compressor tie bolt by a splined coupling. It scavenges oil and entrained air from the turbine rear bearing and returns it to the accessory drive housing. The pump is covered by a thermal insulation blanket and the exhaust inner cone.

Vent System. The air inlet housing cavity and accessory gearbox vent through an external line to a breather mounted on top of the air inlet housing. Seal leakage air from the compressor rear bearing seal vents through the two horizontal struts of the compressor diffuser. The combustor inner casing vents through drilled passages in the turbine coupling shaft and 4th-stage turbine shell into the exhaust gas stream.

START SYSTEM

The engine air start system consists of an air turbine starter, a starter exhaust system, and two independent air supply systems, each having its own control valve. Air from the low pressure starter air control valve enters the starter inlet scroll through a 3-inch line. Air from the high-pressure starter air control valve enters the inlet scroll through a 1.5-inch line. Exhaust air from the starter is discharged through a 6-inch line into the engine module cooling air duct down-stream of the fire damper.

Low-Pressure Air Start System

Air from the ship's bleed air system enters the starter low-pressure air control valve. The control valve is a normally closed, solenoid operated regulating valve that regulates airflow to the starter at 1.83 pounds per second at 45 psig.

High-Pressure Air Start System

Air from the high-pressure air flasks enters the starter high-pressure control valve. The control valve is a normally closed, solenoid operated regulating valve, which regulates airflow to the starter at 2.75 pounds per second at 450 ± 50 psig. A bypass line with an orifice and a rotating pilot valve provides engagement of the starter teeth. A high-pressure start signal causes the pilot valve to open; this allows air to flow through the orifice to the starter at less than 50 psig to engage the teeth. After approximately one-fourth second, the pilot valve opens the air control valve and full pressure (approximately 500 psig) rotates the starter.

Air Starter Motor

The Bendix air turbine starter is mounted on the blind side of the reduction gear high-speed input shaft and drives the engine through the reduction gear during the start cycle.

REDUCTION GEAR AND COUPLING

The reduction gear is a single-reduction, single-helical gear speed reducer. The gear configuration is an over-under, vertically offset, parallel shaft design with a three-piece housing split horizontally at the center lines of the high-speed shaft and the low-speed shaft. The gear elements are supported in sleeve bearings. The starter is mounted on the gear case and drives the high-speed shaft. The oil pump is located at the blind side of the low-speed shaft. The reduction gear is coupled to the generator by a diaphragm flexible coupling. The coupling is attached to the reduction gear and generator shafts by keyed, interference-fit, flanged hubs and can be removed without disturbing either unit, thus preserving alignment.

Lube Oil System

The reduction gear lube oil system is a wet sump, force-feed system. The sump has a capacity of 60 gallons and is an integral part of the reduction gear assembly. It also serves as the supply tank for the gas turbine lube oil system. Oil from the sump is picked up by the reduction gear supply pump rated at 40 gpm at 1800 rpm. From the pump, the oil passes through a 25-micron base-mounted filter, through a remotely mounted oil cooler, and is distributed to the reduction gear, PTO assembly, and to the engine. Pressure at this point is regulated at 25 psig. Oil to the engine and PTO assembly is regulated to 15 psig.

Oil to the PTO assembly is directed by a nozzle onto the shaft coupling and then returned by gravity to the sump. The shaft mid-bearing is lubricated by a spray nozzle on the front of the compressor extension shaft housing. Oil to the reduction gear assembly, 30 gpm at 25 psig, lubricates the reduction gears and bearings and returns by gravity to the sump.

Power Takeoff Assembly

The PTO assembly consists of a PTO shaft, shaft adapter, mid-bearing assembly, housing, and speed sensor pickup. The assembly transmits the torque produced by the engine to the reduction gear. It also provides the means to measure the engine speed with a magnetic pickup over the exciter wheel teeth.

Power Takeoff Shaft and Adapter

The PTO shaft is a solid steel shaft, bolted to the shaft adapter at the forward end and splined to the compressor extension shaft at the aft end. Forty equally spaced teeth are machined on the flange at the forward end of the shaft to provide excitation for the speed sensor.

Housing

The housing encloses the shaft, supports the forward end of the engine, contains the mid-bearing assembly, and provides the mounting for the speed sensor assembly. The mid-bearing assembly prevents the shaft from whipping.

INSTRUMENTATION

The engine assembly incorporates a thermo couple system, a vibration transducer, and a magnetic speed pickup. The thermocouples are wired in parallel to provide an average TIT signal which is amplified by the turbine temperature and speed control in the GTGS control and monitor panel. This signal provides TIT indication and engine emergency shutdown functions.

The signal from the magnetic speed pickup is used by the electronic speed switch in the turbine temperature and speed control for speed sensing and control during startup. An alarm and automatic shutdown are provided in case of an overspeed and under speed. The turbine temperature and speed control also transmit a speed and temperature signal for remote display of engine speed and temperature Digital Display Indicators (DDI).

Thermocouple System

There are 18 dual-element, chromel-alumel thermocouple probes mounted on the turbine inlet casing with the probes extending into the outlet of the combustion liners at the turbine inlet. Each of the probe elements is independent of the other, thus providing two independent sampling circuits. The thermocouple probe housing leading edges are air cooled to prolong probe life. To do this, cooling air enters the probe cavity leading edge through a hole below the probe shoulder, flows through the probe, and is discharged through two small openings in the bottom of the probe.

A thermocouple harness assembly consists of a right and a left section and is enclosed in channels, which are rigidly mounted on the turbine inlet case forward flange. The harness incorporates separate leads for each thermocouple probe. A terminal block serves as the junction for two thermocouple harnesses and the amplifier leads. It consists of eight terminal connections or four terminals for each of the two harnesses.

Vibration Transducer

Engine vibration is measured by a single displacement vibration transducer mounted on the turbine rear bearing support at the 12 o'clock circumferential location.

Speed Pickup

Engine speed is measured by a magnetic pickup mounted in the PTO shaft housing over the shaft exciter teeth. Passage of the exciter teeth under the magnetic pickup produces electrical impulses used by the turbine temperature and speed control for speed sensing; these in turn are used for start sequencing, over- and underspeed protection, and monitoring.

ALTERNATING CURRENT GENERATOR

The generator is of totally enclosed, two bearing construction with an air cooler mounted on top of a fabricated steel frame. It is driven at 1800 rpm from the reduction gear and has a continuous output rating at 50 °C of 2000 kW, 3207 amperes, 450 volts, 3-phase 60 hertz.

Generator Assembly

The generator frame contains internal structural ribs for supporting the stator assembly, baffles for internal air circulation, terminal clocks for internal air circulation, terminal blocks for internal heater and Resistance Temperature Detection (RTD) wiring, and openings for entry of external generator wiring. The frame houses or externally supports the following major generator components:

1. Stator assembly
2. Rotor assembly
3. Front and rear end bracket assemblies
4. Front and rear bearing assemblies
5. Rotor slip ring brush assembly

6. Overhung permanent magnet alternator and lube oil pump assembly
7. Air cooler assembly
8. Stator terminal/connection box

The field rectifier assembly of the exciter/voltage regulator is also mounted in the airstream within the generator enclosure.

Air Cooler

The housing air cooler assembly is mounted in the generator above the generator frame. It is an air-to-water, double-tube, extended-fin heat exchanger. It consists of a core assembly and two water boxes with four zinc anode pencils. The pencils are replaceable units that are inspected periodically for each maintenance procedure. Flanged connections on one water box provide for seawater inlet and outlet. The tubular construction of the core consists of a plain inner tube and an internally fluted outer tube; this tube carries the aluminum cooling fins. In case of a leaking inner tube, the outer tube provides a water passage to the leakage compartment at each end of the core. Each leakage compartment has a telltale space vent and a telltale drain.

Lube Oil System

The generator lube oil system is independent of the gas turbine/ reduction gear lube oil system. The generator system uses mineral oil (2190 TEP) to force-feed the two bearings with 3 gpm at 12 to 15 psig pressure. Oil is taken from the sump tank in the GTGS base by a pump mounted on the permanent magnet alternator shaft, passed through a 25-micron filter and the base-mounted cooler before reading the sleeve bearings. Gravity flow through a sight glass returns the oil to the sump.

Space Heaters

Electric heater elements are provided to prevent the condensation of moisture when the generator is secured or on standby. Four 120-volt,

250-watt, tubular, finned heaters are mounted crosswise under the stator and are spaced to distribute heat along the length of the stator. A heater control switch with an indicator lamp is mounted on the generator switchboard panel. An interlock on the generator circuit breaker automatically disconnects the space heaters when the breaker is closed.

Temperature Monitoring

Nine copper RTDs are imbedded in the generator stator winding slots. The three-wire lead of each RTD is brought to an internal terminal board. A six-position rotary selector switch and a temperature indicator are mounted on the gas turbine generator control panel for monitoring six stator winding temperatures. The three remaining RTDs serve as spares.

A tip-sensitive RTD is embedded in the babbitt of each generator bearing. A terminal assembly, connector, and straight plug are provided for each RTD. A two-position rotary selector switch and temperature indicator, mounted on the gas turbine generator control panel, selects and monitors the two bearing temperatures. Both stator and generator bearing RTD outputs are signal conditioned at the gas turbine generator control panel and transmitted to the Engineering Control and Surveillance System (ECSS) for monitoring.

Generator Excitation and Voltage Control

A static Generator Control Unit (GCU) is associated with each generator and consists of:

1. Static exciter/voltage regulator assembly deck mounted near the associated switchboard

2. Field rectifier assembly mounted in the generator enclosure air path

3. Motor-driven rheostat mounted on the associated switchboard for manual voltage control

4. Mode select rotary switch mounted on the associated switchboard

The GCU provides generator field excitation; maintenance and repair of this equipment is the GSE's responsibility.

GOVERNOR SYSTEM

The engine speed governor is the Woodward 2301 electrohydraulic control system with a backup centrifugal governor override. There are two major components within the system; they are an electronic control unit and an electro-hydraulic actuator. The control unit is mounted in the exciter/voltage regulator enclosure; the actuator is mounted on the gas turbine engine. The control unit is a solid state electronic package that processes input commands and feedback signals to generate a signal to position the actuator. The actuator positions its output shaft in response to the control signal. This shaft controls the engine's liquid fuel valve through a mechanical linkage. If the engine speed increases to a preset limit due to a failure in the electronic control, the centrifugal governor section of the actuator automatically assumes control of the output shaft. Engine speed is then controlled at a point slightly above the normal operating speed.

Operating Modes

The governor system has two basic operating modes—NORMAL (isochronous) and DROOP. The isochronous mode provides constant speed operation, regardless of load. When generators are operated in parallel and in the isochronous mode, the governor system not only maintains a constant speed, but also controls the load division between paralleled generators. Isochronous mode is selected when the Electrical Plant Control Console (EPCC) selector or the switchboard selector is in the NORMAL position. The load-sharing function is automatically enabled when a generator operating in the NORMAL mode is paralleled with another generator.

In the droop mode, the governor system regulates engine speed, but the speed decreases slightly with an increase in load. However, if the generator is paralleled with a constant frequency bus while in the droop mode, the governor cannot control speed, since it is held constant by the bus frequency. Instead, it controls the load carried by the generator. The droop mode is used in this way to provide load control of a generator paralleled with shore power. It also provides a means of unloading a generator paralleled with another generator without disturbing system frequency. Droop mode is selected at the EPCC or the switchboard when the selector is in the DROOP position.

The operating point of the governor is set by a motor operated potentiometer located at the electronic control unit. When the individual frequency adjust controls at the EPCC or the switchboard are enabled, they can be used to adjust the position of the motor-operated potentiometer to a higher or lower position. If generators are operated in parallel with the system frequency controls enabled, the motor-operated potentiometer returns to a calibrated 60-hertz position. Adjustment can be made using the SYSTEM FREQUENCY ADJUST control at EPCC. This control positions a master frequency trimmer in EPCC, which sends equal adjust signals directly to each generator's electronic control unit. In this manner, the frequency of the bus can be changed without disturbing the relationship between operating units. During automatic paralleling operations, the Automatic Paralleling Device (APD) adjusts the oncoming generator for synchronization. This adjust signal is also a direct input into the electronic control unit. It is in effect only during the time of the automatic paralleling conditioning.

Electronic Control Unit

The control unit is modular in design and contains eight major subunits: an amplifier, load sensor, frequency sensor, two power supplies, a motor-operated potentiometer, an accessory box, and two filters.

The motor-operated potentiometer supplies a reference to the amplifier. When the electric plant is operated in the manual or manual permissive modes or with the governor in droop mode, frequency adjust commands cause the motor to rotate in the raise or lower direction, changing the reference correspondingly. When operating in the automatic mode, the motor automatically drives to and remains at a position corresponding to 60 hertz as established by the motor's limit switches. External adjustments to the governor system are then done by additional inputs to the amplifier from either the master frequency trimmer or the APD.

The amplifier provides the current to the actuator. This current is varied in response to the inputs to the amplifier, including the reference, frequency feedback, and load sensing. Changes in the inputs due to load, speed, or reference changes cause the amplifier current to reposition the actuator output shaft, increasing or decreasing fuel flow. Amplifier current stabilizes at a new setting that satisfies all inputs. The amplifier is reverse acting in that the larger the input (error signal), the smaller the output current to the actuator. The actuator output shaft is configured in such a way that a decrease in current causes it to drive the liquid fuel valve toward the maximum fuel position. If the amplifier fails and the current goes to zero, the actuator positions in the maximum fuel position. (The centrifugal governor assumes control if engine speed increases to the preset limit.)

The PMA input to the control unit provides both voltage for the two power supplies and a frequency feedback signal to the frequency sensor. One power supply feeds the amplifier; the second provides power for the motor-operated potentiometer. The frequency sensor converts the PMA output (approximately 120 volts ac at 420 hertz) to a proportional dc voltage for the frequency feedback input to the amplifier.

The load sensor module is used to control load sharing in parallel isochronous operation and to generate the droop characteristics during droop operation. Power generated by the generator is measured by transformers that supply voltage to a bridge circuit. For load sharing, the bridges of each paralleled generator are connected together in a way that an unbalance due to uneven loads causes an input to each governor amplifier. This forces proportional fuel adjustments until the loads are balanced between the two units. This also balances the bridge circuits; the amplifier input is again returned to algebraic zero volts dc. Sudden shifts in load demand cause pulses to develop in the load sensor, which upsets the algebraic zero voltage of the governor amplifier. This results in quicker response to load changes. Polarity of the pulse is also sensed to determine the direction of load changes.

When droop mode is in effect, a portion of the load sensor output is used to oppose the action of the amplifier speed reference. The input to the amplifier decreases by an amount proportional to load; this results in droop.

If the generator is not paralleled with another source, this droop results in a decrease in frequency proportional to the increase in load. If the generator is paralleled with an infinite bus (such as shore power), droop provides load control. When paralleled with an infinite bus, the speed of the machine is held constant by the bus. The governor system cannot then control speed. Any attempt to increase or decrease speed only results in an increase or decrease in load. Without the droop characteristics, the governor system would attempt to adjust the frequency to satisfy the reference exactly. In so doing, it would cause the load to increase beyond capacity or decrease until the flow of power reversed. The droop input, however, modifies the speed reference and the governor reaches a stable operating point even though the frequency does not match the reference. This operating point is set by the speed reference and droop input (since frequency is constant) and determines the load on the generator. Under this condition, the load on the generator remains constant for any reference setting.

Centrifugal Governor

The independent centrifugal governor system is provided as a backup control system to the electronic control unit of the Woodward governor system. If the electronic control unit fails and the engine speed increases due to the actuator positioning itself for full speed failure mode, the centrifugal governor speeder spring device takes over control at approximately 62 hertz (depending on load) or the equivalent speed of approximately14,300 rpm. This is 480 rpm above the 60-hertz speed of 13,820 rpm. The centrifugal governor is part of the hydraulic actuator assembly.

Master Frequency Trimmer

A master frequency trimmer in EPCC provides frequency control to any two or all three generators when operating in parallel.

GAS TURBINE GENERATOR CONTROL SYSTEM

The gas turbine generator control system provides the start/stop sequencing for the GTGS, monitoring and alarms for critical turbine and generator parameters, and signal conditioning for panel meters and transmission of selected data to the ECSS. The gas turbine generator control system is contained in a cabinet mounted on the generator end of the module. On the outside of the cabinet doors are the controls and indicators for local GTGS operation. Inside the cabinet are the electronic components of the system. Among these components are printed circuit cards, voltage regulators, a minus 24-volt dc converter module, a relay assembly, and a temperature and speed control unit. The control elements of the system are powered by 28 volts dc from the switchboard. The switchboard 28-volt dc supply has a bank of 15-ampere hour lead calcium batteries for backup. This battery bank allows starting of a GTGS when the ship is without 450-volt ac power.

Gas Turbine Generator Control Panel

The front doors of the control unit enclosure, mounted on the generator frame, serve as a control and monitor panel and contain the following indicators and meters.

Monitoring and Indicating. A list of these instruments is given below.

1. Generator air temperature
2. Generator bearing temperature (with selector switch to select either bearing)
3. Generator stator temperature (with selector switch to select any one of six temperature detectors)
4. Enclosure temperature (turbine en- closure)
5. High- and low-pressure start cycle counters
6. Lube oil header pressure (reduction gear header)
7. Engine rpm
8. Running time
9. Lube oil temperature (reduction gear header)
10. Turbine inlet temperature
11. Turbine vibration

Shutdown Indications. A list of shutdown indications follows:

1. FAIL TO FIRE—Failure to attain 600 °F TIT within 10 seconds after speed exceeds 2200 rpm

2. START OVERTEMPERATURE—TIT greater than 1600 °F at speeds below 12,780 rpm

3. ENGINE OVERTEMPERATURE—TIT greater than 1945 °F at speeds above 12,780 rpm.

4. ENGINE OVERSPEED—Speed greater than 15,800 rpm

5. UNDERSPEED—Speed below 12,780 rpm after having been above 12,780 rpm for 2 seconds

When engine shutdown occurs due to these abnormal conditions, the control system prevents restarting until the alarm condition no longer exists and the 2-minute coast-down timer has expired.

Alarm Indications Only. Alarm indications are given below:

1. SLOW START—Failure of engine to reach 12,780 rpm within 2 minutes after start initiation

2. HIGH TURBINE INLET TEMPERATURE—TIT greater than 1880°F at speeds above 12,780 rpm

3. EXCESSIVE VIBRATION—Turbine vibrations in excess of three mils peak-to-peak at a frequency above 140 hertz

Other Alarm Displays and Functions. Some other alarm displays and functions on this panel follow:

1. Turbine WATER WASH TANK EMPTY.

2. ALARM SILENCE/ACKNOWLEDGE push button. When an alarm condition occurs, corresponding indicators flash red and the buzzer energizes. The ALARM SILENCE push button also illuminates red. When depressed, the buzzer silences and the indicated

alarm light becomes steady red. When the abnormal condition no longer exists, the alarm indicator turns off automatically.

3. ALARM TEST push button. When de- pressed, this button causes all alarm indicators to flash and the buzzer to sound. The alarms are then reset with the ALARM SILENCE push button. This test assures the operator that the alarm indicators and alarm logic circuits are fully operative.

When the engine is operating with TIT greater than 1600°F and a shutdown occurs due to underspeed, the UNDERSPEED indicator flashes and the START OVERTEMPERATURE indicator illuminates steadily. This condition remains true until the ALARM SILENCE push button depresses and the abnormal conditions no longer exist. This function indicates to the operator the abnormal condition which actually caused the shutdown.

Control Indicators and Push Buttons. A list of control indicators and push buttons follows:

1. D.C. POWER (ON-OFF Toggle Switch)— Provides all power necessary to start and run the gas turbine

2. AC POWER (ON-OFF Toggle Switch)— Provides 115 volts, 60 hertz for enclosure lights and running time meter

3. ENGINE RUNNING Indicator—Illuminates at speeds above 12,780 rpm

4. LOW PRESSURE START (Push Button/Indicator)—Provides for engine starting on the bleed air system (normal)

5. HIGH PRESSURE START (Push Button/Indicator)—Provides for engine starting on the high-pressure system (emergency)

6. STOP—Provides for stopping the engine regardless of the ON-OFF-REMOTE selector switch position

7. MOTOR—Provides for motoring the engine regardless of the ON-OFF-REMOTE selector switch position

8. ON-OFF-REMOTE Selector
   - ON—Enables low- and high-pressure start push buttons on enclosure door
   - OFF—Prevents engine from starting at any location
   - REMOTE—Transfers only start control functions to associated switchboard and/or EPCE (local indicators, alarms, stop, and motor function remain operative).

9. FUEL ENRICH (ON-OFF toggle switch)—Provides for cold weather engine starting. This function has been eliminated as an operational feature.

10. ENGINE ENCLOSURE LIGHT (ON-OFF toggle switch).

11. 14TH-STAGE BLEED AIR (ON-OFF selector switch). ON position enables the control system that opens, regulates, and closes the 14th- stage bleed air valve.

Gas Turbine Generator Control Cabinet

The equipment mounted inside the cabinet consists of the fuses, power supplies, relays, 28-volt dc power line filter, logic cards, signal conditioning cards, alarm cards, solenoid driver cards, contact buffer cards, vibration card, and the temperature and speed control unit.

Both 115 volts ac and 28 volts dc are required for normal operation of the equipment. However, the equipment can start with only the 28-volt dc supply. The GTGS enclosure lights energize through cabinet door interlock relay contacts and the ENGINE ENCLOSURE LIGHT switch. The running time meter energizes through cabinet door interlock relay contacts, the AC POWER switch, and relay contacts which close when the gas turbine is running. All of the 115-volt ac circuits are protected by fuses. The + 28-volt dc electronic circuits are supplied through a power line filter, a 20 amp fuse, and the D.C. POWER switch. The ignition exciter is supplied through a 10 amp fuse and contacts of a relay operated from the logic circuits.

TURBINE TEMPERATURE AND SPEED CONTROL. The turbine temperature and speed control is a combination electronic speed switch and temperature amplifier mounted in the gas turbine generator control cabinet. The control receives a speed signal from a magnetic pickup on the PTO shaft and a temperature signal from the turbine inlet thermocouples. These signals are used to position control relays in four speed channels and five temperature channels within the control, and also provide signals for local and remote monitoring of speed and TIT. In combination with logic circuitry the four speed channels and five temperature channels provide the functions as described below.

2200-RPM Speed Channel:

1. Energizes the ignition circuit.

2. Energizes the fuel shutdown valve solenoid (open).

3. Energizes the fuel pump paralleling valve solenoid (closed) to place the fuel pump high-pressure primary and secondary elements in parallel operation.

4. Deenergizes the start temperature limit valve solenoid, closing the drain port and opening the pressure port to the liquid fuel valve acceleration control section.

5. Arms the start temperature limit valve control circuit to cycle the start temperature limit valve solenoid (on and off) to reduce TIT if it exceeds 1500°F.

6. Starts the 10-second fail-to-fire timer. If TIT fails to reach 600 °F through the 600 °F temperature channel before the timer elapses, the fail-to-fire circuit initiates an automatic shutdown.

7. Energizes the running time meter relay to start the meter.

8400-RPM Speed Channel:

1. Inhibits the fuel paralleling valve, enrichment valve, and ignition circuits to prevent reactivation during a shutdown.

2. Deenergize ignition circuit.

3. Deenergizes the fuel pump paralleling valve solenoid (open) to place the high-pressure fuel pump primary and secondary elements in series operation.

4. Deenergizes the low- or high-pressure start air valve solenoid (closed) to cut out the starter.

5. Extinguishes the low- or high-pressure start push-button indicator.

6. Deenergizes the GTM high-pressure start inhibit relay if the GTGS was started on high- pressure air. Allows GTM in same engine room to be started on high-pressure air.

12,780-RPM Speed Channel:

1. Inhibits the start temperature limit valve control circuit to permit TIT to increase above1500°F.

2. Inhibits the start overtemperature shut-down circuit to permit TIT to increase above 1600 °F.

3. Arms the engine overtemperature shut-down circuit to shut down the turbine if TIT increases to 1945 °F.

4. Inhibits the slow start alarm circuit.

5. Starts the 2-second underspeed timer, which when elapsed arms the underspeed shut-down circuit to automatically shut down the turbine if speed decreases below 12,780 rpm during operation.

6. Illuminates the ENGINE RUNNING indicator light.

7. Enables the 14th-stage bleed valve control circuit to permit opening of the valve (when selected) as long as TIT is not greater than 1850°F.

Once opened below 1850°F, the valve remains open until TIT reaches 1870°F. At this point the valve will start closing to try to hold TIT below 1870°F. When TIT drops below 1870°F, the valve position stabilizes at an interim position. The valve does not close further until TIT again reaches 1870°F and does not completely open again until TIT drops below 1850°F. If TIT remains above 1870°F, the valve closes fully and remains closed until TIT decreases below 1850°F.

15,800-RPM Speed Channel. Provides engine over speed protection. If engine speed exceeds 15,800 rpm, an automatic shutdown is initiated.

600 °F Temperature Channel. Provides automatic shutdown if 600 °F TIT is not reached within 10 seconds after reaching 2200 rpm.

1500°F Temperature Channel. Activates the start temperature limit valve control circuit to intermittently energize the start temperature limit control valve solenoid through a pulse timer to reduce acceleration fuel flow, thereby reducing TIT below 1500°F.

1600°F Temperature Channel. Initiates an automatic shutdown if 1600°F TIT is reached below 12,780 rpm.

1880°F Temperature Channel. Causes an alarm to sound if 18$0°F TIT is reached.

1945 °F Temperature Channel. Initiates an automatic shutdown if 1945 °F TIT is reached above 12,780 rpm.

Automatic Start/Stop

The GTGS control system provides the automatic start sequencing logic, monitors critical parameters, and performs alarm shutdown functions. Following is a description of the events which occur.

START INITIATION. Momentarily press the LOW- or HIGH-PRESSURE START push-button switch. The engine can start from either the low- or high-pressure air system if the respective air systems have been aligned.

The normal start system is low- pressure air; an emergency start system is high-pressure air.

1. The STOP push-button switch will extinguish, and the LOW- (or HIGH-) PRESSURE START push button will illuminate.

2. The low- (or high-) pressure air solenoid will energize. During a high-pressure start of number 1 or number 2 GTGS, the GTM high-pressure start inhibit relay will energize to inhibit the high-pressure start of GTM within the same engine room.

3. The low- (or high-) pressure air start cycle counter will advance one number.

4. The start temperature limit valve solenoid is energized, closing the pressure port and opening the drain, to allow the GTGS to start on the minimum fuel setting.

5. The fuel and ignition circuits are armed.

6. The 2-minute slow start timer is energized.

ACCELERATION UNDER STARTER POWER.—The engine accelerates under starter power and at 2200 rpm the following occurs:

1. The ignition system is energized.

2. The fuel shutdown valve solenoid is energized (opened).

3. The fuel pump paralleling valve solenoid is energized (closed) to place the fuel pump high-pressure primary and secondary elements in parallel operation.

4. The start temperature limit valve solenoid will deenergize, closing the drain port and opening the pressure port to the liquid fuel valve acceleration control section, to accelerate the turbine at a rate equal to CDP increase modified by CIT.

5. The start temperature limit valve control circuit is armed. If TIT reaches 1500 °F, the start temperature limit valve control circuit will intermittently energize the start temperature limit valve solenoid, through a pulse time, to reduce acceleration fuel flow, and thereby reduce TIT below 1500°F.

6. The 10-second fail-to-fire timer is started. If TIT fails to reach 600 °F through the 600 °F temperature channel before the timer elapses, the fail-to-fire circuit will initiate an automatic shutdown. If 600 °F TIT is reached before the timer elapses, the fail-to-fire shutdown circuit is inhibited.

7. The running time meter is energized.

ACCELERATION UNDER STARTER AND ENGINE POWER. The engine accelerates under starter and engine power. At 8400 rpm the following occurs:

1. The fuel paralleling valve, enrichment valve, and ignition circuits are inhibited to prevent reactivation during shutdown.

2. The ignition circuit is deenergized.

3. The fuel pump paralleling valve solenoid is deenergized (open) to place the high-pressure fuel pump primary and secondary elements in series operation.

4. The low- or high-pressure start air valve solenoid is deenergized (closed to cut out the starter).

5. The LOW- or HIGH-PRESSURE START push button/indicator extinguishes.

6. The GTM high-pressure start inhibit relay deenergizes, if GTGS was started on high-pressure air. Allows GTM in same engine room to be started on high-pressure air.

ACCELERATION UNDER ENGINE POWER. The engine accelerates under turbine power and at 12,780 rpm the following occurs:

1. The start temperature limit valve control circuit is inhibited to permit TIT to increase above 1500°F during normal operation.

2. The start overtemperature shutdown circuit is inhibited to permit TIT to increase above 1600°F during operation.

3. The engine overtemperature shutdown circuit is armed, and if TIT increases to 1945 °F, the circuit initiates an automatic shutdown.

4. The slow start alarm circuit is inhibited.

5. The 2-second under speed timer is started, and when elapsed, the under speed shutdown circuit is armed. If turbine speed decreases

below12,780 rpm after the 2-second time has elapsed, an automatic shutdown is initiated.

6. The ENGINE RUNNING indicator light illuminates.

7. The 14th-stage bleed valve control circuit is enabled to permit opening of the valve (when selected) as long as TIT is not greater than 1850°F. Once opened below 1850 °F, the valve remains open until TIT reaches 1870°F. At this point, the valve starts closing to try to hold TIT below 1870°F. When TIT drops below 1870°F, the valve position stabilizes at an interim position. The valve does not close further until TIT again reaches 1870°F and does not completely open again until TIT drops below 1850°F. If TIT remains above 1870°F, the valve fully closes and remains closed until TIT decreases below 1850°F.

The engine continues to accelerate and speed is regulated by the governing system.

ALARM/SHUTDOWN FUNCTIONS. During operation, the GTGS control system provides the logic for the following alarm/ shutdown functions.

1. Alarm Only Functions:
a. SLOW START
b. HIGH TURBINE INLET TEMPERATURE
c. EXCESSIVE VIBRATION

2. Automatic Shutdown and Alarm Functions:
a. FAIL TO FIRE
b. START OVERTEMPERATURE
c. ENGINE OVERTEMPERATURE
d. ENGINE OVERSPEED
e. UNDERSPEED

3. Additional Functions. In addition to the above, the GTGS control system also provides the following functions:

a. Prohibits opening of the 14th-stage bleed air valve at speeds below 12,780 rpm.
b. Inhibits starting for 2 minutes after a shutdown to allow the unit to coast down.
c. Regulates 14th-stage bleed air valve to maintain TIT at less than 1870°F.
d. Provides analog information to the ECSS for alarm and data systems.

Module Gage Panels

Two module gage panels provide continuous indications of important system pressures. The engine gage panel is located on the module enclosure side at the generator end. The gearbox gage panel is mounted on the gearbox. It is visible only when the gear enclosure door is open. The gages on these panels are mechanical gages piped directly to the sensing point.

Sea Water Service System

Each GTGS has an independent seawater service system for the lube oil coolers and generator air cooler. An electric pump starts automatically as the generator voltage builds up since power is taken from the generator side of the main circuit breaker. If the electric pump system fails, emergency cooling water is supplied by the ship's seawater service system. A solenoid-operated pilot valve opens and automatically opens the diaphragm-activated, hydraulic-operated stop valve when low pressure contacts close in the pressure switch in the normal service line. Cooling water is drawn from the sea chest and flows through the generator air cooler and lube oil cooler, and then through the engine lube oil cooler. As the seawater flow requirements are different for the three coolers, each unit has a bypass valve to adjust seawater flowing through the cooler.

Waste-Heat Boiler

Steam for ship's services is generated by three waste-heat boilers using the hot exhaust gas from the GTGSs as the heat source. The boilers are located in numbers 1 and 2 engine rooms and in the number 3 waste-heat boiler room located above the number 3 generator room.

Numbers 1 and 2 waste-heat boilers are identical. The number 3 boiler is identical except for minor piping differences which are necessary to facilitate positioning of the operating controls and control panel.

The boilers are of the forced-recirculation, water-tube type. Recirculation of the boiler water is provided by a high head recirculating pump which delivers a minimum of 500 percent excess water under maximum evaporation demand. The boiler tubes are finned and arranged in a coiled bundle configuration. Exhaust gas enters the bottom of the boiler and is discharged from the side of the boiler casing.

Rated steam capacity of each boiler is 7000 pounds per hour at a nominal operating pressure of 100 psig and a gas inlet temperature of 650 °F. Detailed design capabilities and limitations are shown in table 4-1. The system consists of the following equipment:

1. Boiler
2. Steam separator/water reservoir
3. Recirculating pump
4. Control condenser
5. Controls
6. Miscellaneous components
7. Boiler control panel
8. Condensate system
9. Feed water and drain collecting system

Table 4-1.—Boiler Design Characteristics

PARAMETERS	DESIGN CHARACTERISTICS
Rated Steam Output	7000 lb/hr
Design Pressure	120 psig
Maximum Operating Pressure	115 psig
Nominal Operating Pressure	100 psig
Turbine Exhaust Gas Temperature In	1035°F maximum
Boiler Exhaust Gas Temperature Out	450°F maximum
Turbine Exhaust Mass Flow	34.2 lb/sec
Safety Valve Lifting Pressure	126 and 130 psig
Safety Valve Reseating Pressure	108 and 112 psig
Feedwater Temperature	130°F minimum

The Gas Outlet Temperature shall not exceed 700°F except during dry running. If the Gas Outlet Temperature rises to 700°F, the boiler should be secured and the fin surfaces cleaned.

Boiler

The boiler consists of a three-piece, gas-tight casing, the steam generating coils or tube bundle, and four soot blowers. Hot exhaust gas from the ship's service generator gas turbine engines enters the boiler from the bottom, passes around and between the boiler coils heating the water inside the tubes, and discharges through the side of the casing. The casing, consisting of a half section and two quarter sections, is fabricated with an inner wall of 10-gauge steel, an outer wall of steel, and a 4-inch thick blanket of spun glass insulation sandwiched in between. Access ports are provided for maintenance and inspection.

Tube Bundle

The boiler coil or tube bundle is made up of twenty-seven 1.25-inch outside diameter, smooth bore, finned, alloy steel tubes. The tubes are wound in flat spiral coils and are assembled to the inlet and outlet headers so that the spaces between coils coincide. This arrangement allows the flow of gas turbine exhaust gas around and between the tubes for maximum heat transfer with minimum pressure drop across the boiler.

When a leak develops in a waste-heat boiler, isolating the leaking tube is difficult. To positively identify the leaky tube a method has been developed to perform individual tube hydrostatic tests. The outlet header must be modified by installing finger holes which allow you to plug the tube end with a special plug provided as part of the kit. The opposite end of the tube is plugged with another plug which allows a vacuum gun to be hooked to it. Ship's service air is sent through a nozzle in the gun. The nozzle draws approximately 20 inches vacuum on the tube. A gage mounted on the gun measures the vacuum. If the vacuum holds the tube is good, and if it falls to zero the tube is leaking. This system allows you to quickly identify a leaking tube and ensure the correct tube is plugged if necessary.

Steam Separator/Water Reservoir

The steam separator/water reservoir is vertically mounted to the base adjacent to the boiler. It consists of a steam drum or flash chamber containing a cyclone centrifugal steam separator, a dry pipe, and steam baffles. Steam separation takes place in the top of the unit with the steam being discharged into the dry pipe connected to the top of the separator and the water being discharged into the lower part. In operation, a mixture of steam and water is discharged from the boiler into the steam separator inlet connection, which is at a tangent to the separator body. The mixture then passes through an orifice within the inlet connection, which causes it to increase in velocity as it enters the separator. Because of the angle of entrance, increased speed, and helix device within the separator, the mixture acquires a rotary motion. Due to the rotary motion of the mixture, centrifugal force separates the water from the steam. The water, being heavier, is thrown to the sides while the lighter steam remains in the center. An internal baffle further deflects steam to the center and water to the sides. The steam then rises through a scrubber element, which causes the steam to change flow direction several times with further removal of moisture. From the scrubber element, the steam passes through an orifice and enters the dry pipe which carries it to the steam header. The separated water falls to the base of the separator. Vanes in the base of the separator serve to maintain the rotary motion of the water until it is discharged from the outer edge of the separator into the reservoir. Connections are provided on the steam drum for high- and low-water level control, feed- water inlet, boiler water outlet, and a water level sight gauge. A manhole provides access for inspection and maintenance. Provision is also made for bottom blowdown and draining.

Recirculation Pump

The purpose of the recirculation pump is to provide a continuous flow of water from the steam separator/water reservoir to the tube bundle and back to the steam separator. The pump is a horizontally mounted, single-stage, centrifugal type, capable of recirculating boiler water at a rate which exceeds the evaporation rate. It delivers up to 144 gpm at 30 psid. Water piped from the feedwater inlet line to the pump seal cools and lubricates the mechanical seal assembly. The pump is driven by a constant speed induction motor rated at 5 horsepower.

Control Condenser

The purpose of the control condenser is to condense the excess steam resulting from variations of steam demand and turbine exhaust gas temperature. The condenser is a shell/tube design with the coolant (seawater) flowing through the tubes. It has the capacity to condense the total boiler output at all operating conditions. Seawater is provided to the condenser at 50 psig within a temperature range of 28 °F to 85 °F. Maximum seawater outlet temperature is 140°F. A pneumatically operated valve in the seawater outlet piping and a pneumatic temperature controller regulate the flow of seawater to prevent the temperature of the condensate output from exceeding 170 °F.

Steam Pressure Control

Steam pressure control is accomplished through control of the steam dump valve. Steam is discharged from the steam separator into a dry pipe. The steam manifold splits into two lines at a "Y" connection. One line goes to the steam distribution header; the other line goes to the control condenser. The steam stop valve is located in the line to the distribution header; the steam dump valve is located in the line to the condenser. Both valves are 4-inch, cast-steel, single-seated, diaphragm- actuated control valves. The steam stop valve is normally closed, air to open; the steam dump valve is normally opened, air to close.

Steam Stop Valve

The steam stop valve is either open or shut. It is actuated by control air from the ship's service air system reduced to 60 psig by a reducing valve. A solenoid valve, when energized, admits control air at 60 psig to the steam stop valve actuator, opening the valve. To deenergize the solenoid valve blocks the 60 psig control air pressure and vents the stop valve actuator to atmosphere, then the stop valve closes.

Steam Dump Valve

The steam dump valve is modulated to control the amount of steam diverted to the control condenser. The steam pressure controller senses steam pressure and regulates control air (ship's service air reduced to 30 psig by a reducing valve) to position the dump valve. An increase of steam pressure causes a decrease of regulated air pressure and the dump valve tends to open. This diverts steam to the control condenser, thus reducing steam pressure. A decrease of steam pressure causes an increase of regulated air pressure and the dump valve tends to close; this increases steam pressure. The steam dump valve can also be manually operated.

An emergency stop switch is provided at the waste-heat boiler control panel and at the PAMCE. Depressing the emergency stop switch, deenergizes the solenoid valve associated with the steam stop valve; this closes the stop valve. A waste-heat boiler summary alarm and a gauge to monitor forward and aft steam header pressures is also provided at PAMCE.

The boiler is protected against overpressurization by two safety valves, vertically mounted on the shoulder flanges of the steam separator drum. The valves are identical except for their pressure settings. One valve is set to lift at 126 psig, the other at 130 psig. Steam discharge from the safety valves is piped to the exhaust gas outlet duct.

Feedwater Control Valve

The feedwater control valve is a 1-inch, cast- steel, normally closed, air-to-open, diaphragm- actuated control valve. It is located in the feedwater line upstream of the chemical tank. Valve control is automatic with a handwheel provided for manual operation in the event of failure.

The feedwater control valve is modulated by the boiler water level controller, to regulate the water level in the boiler (steam separator/ reservoir). The boiler water level controller senses the water level in the steam separator/reservoir as a differential pressure and provides a variable air pressure (from the 30 psig reducer) to the feed- water control valve actuator to control flow. When the water level rises, air pressure reduces and the feed water valve tends to close. When the water level drops, air pressure rises and the valve tends to open.

There are two solenoid actuated valves in the air line to the feed water control valve. They are the high-water solenoid valve and the low-water solenoid valve. Both valves are open (energized) during operation.

When low boiler water level is sensed, an indicator light on the control panel illuminates. If the water level remains low for 2 minutes, an alarm sounds at PACC and the low boiler water level solenoid valve is deenergized by the 27 psi pressure switch bypassing the boiler level controller and porting full low pressure air (30 psig) to the feedwater control valve actuator. The feedwater control valve then goes fully open. When high boiler water level is sensed, an indicating light on the control panel illuminates, the high boiler water level solenoid valve is deenergized by a pressure switch sensing 12 psig control pressure and blocks control air pressure to the feedwater control valve actuator, which vents the actuator to the atmosphere. The feedwater control valve then goes fully closed. There is only a local indicator light associated with high boiler water level. A low boiler water level activates the respective boiler summary fault alarm at PAMCE.

MISCELLANEOUS COMPONENTS

Some additional components that support the boiler operation are the chemical treatment tank and the blowdown and soot blower system.

Chemical Treatment Tank

The chemical treatment tank provides for chemical treatment of the boiler feedwater. Parameters for boiler water chemistry are provided in Naval Ships' Technical Manual (NSTM), chapter 220. The tank is located in a branch line off the feedwater line downstream of the feedwater control valve. The tank has a liquid capacity slightly greater than 1 gallon. Water treatment chemicals are released into the tank through a funnel on the top at the tank inlet valve. To add chemical treatment to the feedwater, the feedwater globe valve is closed and the feedwater chemical inlet and outlet valves are opened. This diverts the feedwater through the chemical tank before it is delivered to the boiler.

Some ships have a continuous chemical injection system installed. It consists of a chemical injection tank which is prefilled with a solution of trisodium phosphate and disodium phosphate which is injected into the system by a proportioning pump. The pump stroke can be varied to provide proper mixture of chemicals to maintain boiler chemical balance. NSTM, chapter 220 provides further information on this system.

Blowdown

Blowdown of the steam separator and the inlet and outlet headers is used for the removal of sludge and sediment. The blowdown is provided for through an outlet on the bottom of the separator and through valves at the low point of each header. The boiler blowdown is piped overboard through the hull and through a tee connection to a funnel, which leads to the waste oil drain tank in the main engine rooms and to the bilge in the number 3 generator room. The blowdown valves are manually operated. This piping is also used to drain the boiler.

Soot Blowers

The soot blowers consist of four stationary steam lances positioned vertically between the coils, 90 degrees apart, and are controlled by individual valves. Steam, applied at approximately 100 psig, discharges from holes in the lance arranged so that the steam covers all sides of the tubes. Soot, carbon, and combustion gas residue is discharged through the exhaust gas outlet.

BOILER CONTROL PANEL

The control panel is an integral part of the waste-heat boiler. It contains the boiler controls, warning lights, an alarm bell, indicating lights, and gauges for local monitoring of all boiler operating parameters and steam distribution pressure.

Remote Control and Monitoring

An emergency stop switch is provided at PAMCE which, when depressed, automatically closes the steam stop valve.

Warning lights are provided at PAMCE for the following faults:

1. Low Steam Header Pressure:
    - FWD HEADER PRESS LO 90 psig
    - AFT HEADER PRESS LO 90 psig

2. Presence of Oily Condensate: OILY CONDENSATE 25 ppm

3. High Salinity: CONDENSATE COOLER NO. 1 SALINITY HI .05 epm

4. High/Low Feedwater Tank Level:
    - FWD FEEDWTR TANK LEVEL HI/LO— 471/90 gallons
    - AFT FEEDWTR TANK LEVEL HI/LO— 526-108 gallons

A summary fault light is provided for the following faults:

1. Low control air pressure.
2. Low boiler water level.
3. High or low steam pressure.

A gauge is also provided at PAMCE for continuous monitoring of steam header pressure.

CONDENSATE COLLECTING SYSTEM

Condensate from the control condensers, distilling plants, and hot water heating system is returned to the condensate main. From the condensate main, the condensate then passes through the condensate cooler to the feedwater and drain collection tank. The condensate that passes from the control condenser to the condensate main is monitored for salinity. The condensate that passes from the condensate cooler to the feedwater and drain collecting tank is also monitored for salinity.

Condensate from heaters that interface with fuel oil or lube oil is returned to the oily condensate main. From the oily condensate main, the condensate passes through the contaminated drain inspection tank where it is monitored for the presence of oil, and then to the condensate main where it joins the condensates mentioned in the previous paragraph. The feedwater pump draws condensate from the feedwater and drain collecting tank and pumps it back to the feedwater control valve as boiler makeup water.

Contaminated Drain Inspection

Condensate from the fuel and lube oil heating systems is monitored for the presence of oil contamination by way of two contaminated drain inspection tanks before it is returned to the feed and drain collecting system. The inspection tanks are installed in the contaminated drain piping near respective feed and drain collecting tanks in numbers 1 and 2 engine rooms.

The inspection tanks are equipped with ultraviolet oil detectors that continuously and automatically monitor the condensate for oil contamination. The oil detector consists of a sample assembly installed in the tank inlet piping and a bulkhead mounted control box assembly. The sample assembly consists of a sample chamber, an ultraviolet lamp, and an ultraviolet sensor. All the condensate from the contaminated drain system passes through the sample chamber where ultraviolet energy transmitted through the stream is measured by the sensor. Clean, oil-free water absorbs a small amount of ultraviolet energy; however, when even small amounts of oil are mixed with the water, either in the form of droplets or as an emulsion, a reduction in ultraviolet energy transmission results, which is readily detected by the sensor. The control box contains the electrical control circuit and alarm and warning relays. The front panel contains the ON/OFF switch and green NORMAL and red ALARM status lamps. The inner panel, contains the percent meter, adjustment controls, and test switches.

When oil (warning level) is detected in the contaminated drain system, the warning relay deenergizes and the forward unit illuminates the DRAIN CONTAMINATION indicator light on the number 1 boiler control panel. The after unit illuminates the DRAIN CONTAMINATION light on the number 2 boiler control panel in a similar manner. The DRAIN CONTAMINATION light on the number 3 boiler is not connected. When oil at the alarm level is detected (25 ppm set point), the alarm relay deenergizes and the associated solenoid-operated valve diverts the contaminated condensate to the waste oil drain system. At the same time, the NORMAL light is extinguished and the ALARM light is illuminated on the control box front panel. In addition, the control box sends an alarm signal to the ECSS to illuminate the OIL CONDENSATE summary alarm indicator light at PAMCE and to provide alarm data to the alarm log.

The alarm and warning circuits remain activated as long as contamination remains above the set point. However, when contamination drops below the set point, the alarm and warning indicator lights extinguish and the solenoid-operated valve closes.

The contaminated drain tank is equipped with a sight glass for visual inspection.

Condensate Cooler

The condensate cooler is a shell/tube design with the coolant (seawater) flowing through the tubes. A pneumatically operated valve in the seawater outlet piping and a pneumatic temperature controller regulate the flow of seawater to regulate the temperature of the condensate output at 140°F on systems without a deaerating feed tank (DFT) system and 120°F for systems with a DFT. When feed water temperature exceeds 170°F it causes pump cavitation.

SALINITY INDICATION SYSTEM

There are three independent salinity indicating systems monitoring the condensate return system, one system associated with each waste-heat boiler. The two forward systems are similar and each consists of a bell and dump module, two salinity indicator and controllers, and two salinity cells monitoring the control condenser and condensate cooler drains for the associated waste-heat boiler (numbers 1 and 2). The system for monitoring the control condenser drain of waste-heat boiler number 3 consists of one salinity indicator and controller, and one salinity cell.

Each salinity indicator and control panel contains a yellow POWER ON light, a CHLORIDE EPM meter, a red ALARM light, a red BELL CUTOUT light, and a TEST METER ALARM switch. The internal electronic circuits provide a continuous salinity reading for the meter as well as deenergizing a relay when salinity exceeds a set point.

The POWER ON light indicates 115 volts ac is present in the controller. The CHLORIDE EPM meter gives a continuous reading of salinity as sensed by the salinity cell. When salinity exceeds the set point, a relay deenergizes to illuminate the ALARM light and also energizes the alarm bell on the bell and dump module. Operation of the SILENCE switch silences the alarm bell and illuminates the BELL CUTOUT light. The TEST METER ALARM switch provides a means of checking the system alarm by substituting a resistor for the salinity cell.

The bell and dump module panel contains an alarm bell, a red BYPASS ALARM light, a green SOLENOID light, and a BYPASS switch. Operation of the BYPASS switch energizes the dump valve solenoid to prevent it from dumping and also illuminates the BYPASS ALARM light. The SOLENOID light is normally illuminated steady. If, however, a high salinity signal is received from the condensate cooler salinity indicator and controller, the SOLENOID light will start flashing. It continues to flash until the salinity value at the condensate cooler falls below the set point, regardless of the position of the dump valve.

The two forward systems operate in the same manner. When a high salinity signal is received from the control condenser salinity indicator and controller, a relay deenergizes in the bell and dump module. This relay illuminates the red SALINITY light on the waste-heat boiler control panel (numbers 1 and 2), sends a signal to PAMCE to illuminate the CONDENSATE COOLER number 1 or 2 SALINITY HI light, and also provides data for the alarm log.

When a high salinity signal is received from the condensate cooler salinity indicator and controller, two relays deenergize in the bell and dump module. The first relay is the same one as described in the previous paragraph. The second relay deenergizes the dump valve solenoid to divert condensate to drain. It also causes the SOLENOID light to start flashing.

The salinity system for waste-heat boiler number 3 is a local system and consists of one salinity indicator and controller and one salinity cell. The salinity cell is located in the control condenser drain. In the event of high salinity, the salinity indicator and controller illuminates the red ALARM light on the controller as well as the red SALINITY light on the number 3 boiler control panel.

Feedwater and Drain Collecting System

The feedwater and drain collecting system consists of feed tanks located in the numbers 1 and 2 engine rooms and are cross- connected. Makeup water is supplied to the tanks from the distilling plant. Additionally, provision is made through an air gap (funnel) to the freshwater system for an emergency supply. The tanks are designated as the feedwater and reserve feedwater tanks dependent upon the plant setup as designated by the Chief Engineer.

Tank level is automatically controlled by a float-type level controller. When the water level in the tank drops below a set level, the makeup feedwater solenoid valve opens and makeup water flows into the tank. When the tank is full, the makeup valve closes, and shuts off the makeup water.

The level controller also contains switches for high and low water level indicating lights located on the boiler control panel and at PAMCE.

Feedwater Pumps

The purpose of the feedwater pumps is to supply feedwater to the boiler to maintain the proper water level in the reservoir during boiler operation. There are three feedwater pumps: the number 1 pump located in the number 1 engine room and the numbers 2 and 3 pumps located in the number 2 engine room. The feedwater piping system is such that feedwater can be cross-connected. The feedwater pumps are of the single stage, horizontally mounted, centrifugal type. The pumps are driven by constant speed induction motors, rated at 20 horsepower. The rated capacity of each pump is 52 gpm at 130 psig.

Deaerating Feed Tank

An addition to the waste heat-boiler that is being installed on all ships at the present time is the DFT system. The purpose of this system is the removal of gases in the feedwater which, in the case of oxygen, leads to corrosion and pitting of the internal surfaces of waste-heat boiler tubes and other equipment. In waste-heat boilers this pitting eventually leads to failure and leakage of boiler tubes. By removing dissolved oxygen in the feedwater with the DFT the reliability and life of the waste-heat boiler system is increased.

The feedwater booster pump takes feedwater from the ships feedwater tanks and directs it to the sub-cooler. This feedwater is used to cool the water from the DFT prior to entering the waste-heat boiler feed pump and prevent it from flashing to steam in the waste-heat boiler feed pump. Feed- water leaves the sub-cooler and flows through the feedwater control valve. The feedwater control valve is an air operated valve and controls the flow of feedwater to the DFT based on the level of the DFT. It is used to maintain normal water level in the DFT. A spillover valve is also used to prevent a high water level from occurring and to maintain a minimum feedwater flow of 10 gpm. Feedwater flow from the spillover valve is returned to the feedtank through the condensate cooler.

Feedwater enters the DFT through a spray valve located in the spray head. The spray valve discharges the feedwater in a fine spray into the steam-filled upper section of the DFT. The extremely small droplets of water are heated and partly deaerated by the relatively air-free steam. As the steam loses its heat to the water, much of the steam is condensed to water. The droplets of water (including both the entering feedwater and the condensed steam) are joined in a conical collection chamber and directed to the deaerating section.

Steam enters through the atomizing valve, picks up the partially deaerated water, and throws it outward against the baffles and the deflector plates of the deaerating unit. In this process the liquid is even more finely divided and is thoroughly scrubbed by the incoming steam. Thus, the last traces of dissolved oxygen are removed from the water. Since the water enters the deaerating section at saturation temperature, having already been heated by the steam in the upper part of the tank, the incoming steam does not condense to any marked degree in the deaerating section. Therefore, practically all of the incoming steam is available for breaking up, scrubbing, and deaerating the water.

The deaerated water drops into the storage space at the bottom of the tank where it remains until it is evacuated by the boiler feed pump. Meanwhile, the mixture of steam plus air and other noncondensable gases travels up and through the feedwater spray (where much of the steam is condensed, as it heats the incoming water). The steam-gas mixture then exits the tank and flows to the external vent condenser. The remaining steam is condensed in the vent condenser and the water is returned to the feedwater holding tank. The air and other noncondensable gases are vented to the atmosphere from the shell of the vent condenser. The DFT feedwater is then pumped to the waste-heat boiler by the boiler feedpump.

Steam pressure is regulated to 17 psig within the DFT and a relief valve is set to relieve the internal pressure of the DFT at 30 psig.

Summary

The Allison gas turbine engine provides the prime mover to the ship's service generator on the DDG and CG class ships. The model 104 which you have studied in this chapter is the most common unit found on gas turbine ships. Associated with the engine is a waste-heat boiler that uses the exhaust gases from the engine passing around water-filled tubes of the boiler to provide steam for the ship.

You have seen how the engine is constructed, how it functions, and how it is controlled. The interface between the boiler and engine has been described. The generator and its operation have also been discussed. You, as an engineer, should now be familiar enough with the system components to be able to follow technical manual instructions and maintenance procedures. The GTGS and waste-heat boiler is an important part of the overall engineering plant. With the knowledge gained in this chapter, you should be able to help maintain and repair this equipment.

Detailed instructions for starting, operating, securing the boiler, and laying it up are found in the EOSS, Naval Ships' Technical Manual, and ship's instructions and should be referred to for proper procedures.

CHAPTER 5

GAS TURBINE INSPECTION AND TROUBLESHOOTING

In this chapter we will discuss object damage, borescope inspection, and troubleshooting of the LM2500 GTE. We will also briefly discuss the borescope inspection and troubleshooting of the Allison 501-K17 GTE. The majority of this chapter deals with the LM2500 GTE damage evaluation, accompanied by numerous explanations and terms used in the text.

Object Damage

One of the most damaging casualties to a gas turbine, and one of the easiest to prevent, is Foreign Object Damage (FOD). In this section we will discuss the hazards of FOD and some of the ways to prevent it. Another type of object damage that can cause failure of a GTE is Domestic Object Damage (DOD). This type of damage occurs when an internal object from the engine breaks loose and causes impact damage to the engine.

Hazards

The effects of FOD and the hazards involved vary greatly with the size of the object ingested. Small dents and abrasions may cause little or no damage; however, if a large enough object is ingested by the engine, severe internal damage will result. Large, soft items (such as paper) can clot the FOD screen, causing a loss of power and elevated turbine inlet temperatures.

Prevention

To prevent FOD to engines while working in and around intake and plenum areas, you must observe the following safety precautions:

• When performing maintenance inside the intake areas, remove all loose objects from your person. You must also account for all tools and equipment used in the intake. After completing your work, inspect the intake for cleanliness, and reinventory the tools and equipment before securing the accesses.

• Periodically inspect all intakes for cleanliness, the state of preservation, and the condition of the FOD screens. Correct any abnormal conditions. The frequency of inspection will depend on the operating conditions, PMS requirements, and engineering department instructions.

• When inspecting the intakes, ensure that the areas around the blow-in doors are kept clear of loose gear and debris that could be ingested if the blow-in doors are activated.

BORESCOPE INSPECTIONS

You will find the borescope inspection requirements and procedures on the MRC. These cards contain all the basic information necessary to conduct the inspection. Included on the MRC is a list of conditions that would require an inspection and the serviceability limits. Borescope inspections are done semiannually or when the engine has been operated beyond the allowable limits as listed on the MRC.

In the following section we will discuss the borescope procedures used to inspect the LM2500 GTE. You may also apply the inspection procedures and the knowledge gained from damage evaluation to the borescope inspection of the Allison 501-K17/34 GTE.

General Inspection Procedures

It is good engineering practice to review the history of an engine before you conduct an inspection. Various component improvement programs will eventually affect all engines in service. A rebuilt or modified engine may contain improved parts that are different in appearance from the original. An example of this is the first-stage compressor mid-span damper that may have its original coating, an improved coating, or a Carboloy shoe welded on at the mid-span damper interface. If you review the machinery history, you will know the status of the parts that have been changed or modified.

Assuming that the engine history is normal and FOD is not suspected, you should be aware of the following factors when conducting a borescope inspection.

- Know your equipment.
- Locate all inspection areas and ports.
- Establish internal reference points.
- Scan the inspection area thoroughly and in an orderly manner.
- Note any inconsistencies.
- Evaluate the inconsistencies.
- Report your conclusions.

Geometric Orientation of the Engine

When the probe is in the inspection hole, it is not unusual for you to lose your sense of direction. On the Wolf borescope, the large plastic disk just beneath the eyepiece has an index mark that shows the direction the probe object window is facing. You can feel and see this mark. Another reference you can use to detect the direction the object window is facing is the light cable attachment. On the Wolf and Eder probes, the viewing window is 90 degrees clockwise from the light cable. Later models of borescoping equipment may have changes incorporated that improve the equipment. Newer models may incorporate a swiveling light cable that allows the cable to hang down regardless of the viewing direction. You must read the manufacturer's instruction manual before using the equipment.

Borescope Ports

Compressor

Fifteen borescope inspection ports are in the compressor near the 3 o'clock split line. A port is located at every compressor stator stage. These vane ports start at the IGVs and work aft in the same direction as the airflow (except for stage 8, which is internally blocked). Stator stages 9 and 13 borescope ports require you to remove piping interferences.

Combustor and HP Turbine

Aft of the right-hand side compressor ports are six circumferentially positioned ports, just forward of the midflange of the compressor rear frame. From these ports you can inspect the combustor, the stage 1 HP turbine nozzle assembly, and a few fuel nozzles. Near the aft flange of the compressor rear frame on the right-hand side of the engine are two HP turbine stator ports that you can use for viewing the air-cooled turbine blades. The Pr5.4 pressure probe harness adjacent to the after flange of the turbine mid frame is located aft of the stage 1 and 2 turbine ports. Five pressure probes are located circumferentially around the turbine mid frame at the inlet to the LP turbine. All five probes extend radially into the gas path and can be removed to inspect the LP turbine inlet and the HP turbine exhaust.

INDEXING AND ROTATING THE ENGINE

You can rotate the engine by using a socket wrench with an 18-inch long 3/4-inch drive extension. The 3/4-inch drive is installed after you remove the cover plate on the aft face, right-hand side of the accessory gearbox, next to the lube and scavenge pump. When you are inspecting through the forward-most borescope ports, there is not enough space for both you and the person turning the engine to work. This requires you to do the turning yourself or to have the turner rotate the engine from the other location on the accessory gearbox. You can find the alternate drive pad for manual engine turning on the forward face, left-hand side of the accessory gearbox.

Detailed procedures are provided for indexing and rotating the engine on the MRC. Zero reference for the compressor and HP turbine stages is established by use of the locking lug blades. Establishing the zero reference ensures a complete inspection for each stage. It also provides you an immediate circumferential reference point for distress reporting for each stage. You should not concentrate on counting the blades. Instead, concentrate on the specific condition of each airfoil as it passes. The forward and aft drive pads have different drive ratios to the main rotor shaft. You may find it advantageous to use a torque multiplier to slow down and maintain better control over the main rotor speed. Depending on the manual drive setup, you will be able to establish how many full arcs of the ratchet wrench are required to move the main rotor one full revolution. For example, when you are using the forward pad, a 344-degree revolution of the input drive equals 360 degrees on the main rotor.

SERVICE LIMITS

In this section we will discuss the types of damage that you might find when conducting a routine inspection. This material will be limited to a discussion of the major engine areas.

Compressor Section

You should inspect the compressor section for nicks and dents, cracks, spacer rubs, casing rubs, blade tip rubs, bent edges, missing pieces, and trailing edge erosion. Inspect the first-stage compressor mid-span damper for leading edge dents and other types of damage. Beginning with the third stage, if a slight tilting of the blade or raising of the blade platform is observed, suspect blade root failure. This condition requires suspended engine operation until the condition has been evaluated.

COMPRESSOR DAMAGE. In the following paragraphs, we describe some of the damage you may find during an engine inspection.

Airfoil and Tip Cracks. Cracks in the compressor hardware are difficult to detect because they are tight and shallow in depth. You can miss these subtle defects because of deteriorated borescope optics or if you rotate the rotor too fast. You should record all crack information relative to the stage, area, magnitude, direction, and adjacent blade condition.

Cracked Dovetail. A cracked dovetail of a blade may lead to blade loss. The location of the blade will determine the extent of engine damage. Before the actual catastrophic failure of the blade, the separated crack in the dovetail will be evident by a leaning blade platform. You can find this fault by using the borescope to inspect each blade platform. The leaning blade platform will be higher than the adjacent nonleaning blades. A "leaner" is a blade that has a crack on the aft side of the dovetail and is leaning in the forward direction. If a leaner is detected, it must be verified and the engine should be removed from service.

Airfoil and Tip Tears. The most critical area of a torn blade is the area around the end of the tear and its location on the airfoil. You should inspect this area for cracks that lead from the tear and are susceptible to propagation. This condition could lead to the loss of the airfoil section that would create downstream impact damage. You should record all information such as stage, blade locations, area of the blade in which the defect was found, and the condition of the rest of the airfoil and adjacent airfoils.

Leading and Trailing Edge Damage. Random impact damage can be caused throughout the compressor rotor stages by FOD and DOD. The leading and trailing edge of an airfoil is the area of the compressor blade extending from the edge into the airfoil. You must assess both sides or faces of the airfoil when determining the extent of a given defect. If you observe a defect, estimate the percentage of damaged chord length. Observe the defect and the condition of the airfoil area around the defect. If the observed damage is assessed to be "object damage," the most difficult determination is the differentiation between cracks, scratches, and marks made by the passing objects. Cracks are usually tight in the airfoils, but the apex of the damage usually allows viewing into the airfoil thickness. This provides a direct inspection of the area around the crack. You may have to use all the probes at varying light levels to determine the extent of the damage.

Tip Curl. Compressor rotor blade tip curl is a random and infrequent observation. Tip curl is usually the result of blade rub on the compressor case. Tip curl can also be the result of objects being thrown to the outer circumferential area of the flow path and then being impacted by the rotating blade tip (either leading or trailing edge). These curled tips are usually smooth in the bend area of the airfoil distortion; however, you should inspect the area at the change in normal airfoil for tears or cracks. When you report tip curl, estimate the percent of the chord length, the number of blades with curl, and the condition in the adjacent airfoil area. Record any evidence of impact and inspect for the origin of the impact. Always look at the adjacent blades for evidence of tip clang.

Missing Metal. Missing metal from compressor rotor blade airfoils is a result of the progression of cracked or torn airfoils that release part of the airfoil into the flow path. Crack propagation in the root fillet area can result in the separation of the entire blade. Severe FOD or DOD may result in several random rotor and stator airfoils with missing metal. The inspection report should include the stage, the number of blades with missing metal, the amount, and the location on the airfoil. Estimate the percent of chord, the span of the airfoil that is missing metal, and the condition of the remaining airfoil.

Airfoil Surface Defects. Surface defects are the result of object damage or adjacent blade interference (tip clang). Impacts in the center section of the airfoil are not common. Tip clang damage is the result of a blade leading edge tip contacting the adjacent blade tip at approximately one-third of the chord length forward of the trailing edge on the low-pressure (convex) side of the blade. This is the result of compressor stall and is observed in stages 3 through 6. You should report an observed defect on the airfoil surfaces in the inspection record. Your report should contain information relative to the stage, location on the blade (estimate the percent of chord and span), and the condition of the surrounding airfoil. You do not have to record the appearance of the defect (sharpness and contour). Compressor stall is one of the worst things that can happen to an engine. Tip clang damage is difficult to spot and gives the appearance of minor damage. The V-shaped notch on the top of a blade caused by tip clang is only an indicator; it in itself is not the damage. The damage is at the blade root and normally cannot be seen. If the blades have been over-stressed, they must be replaced.

Platform Distortion. Compressor blade platform shingling can be observed on some after stage blades. Shingling is the over- lapping of one blade platform mating edge with the adjacent platform edge. When shingling is found, the platforms will be distorted and bowed. When the platforms are shingled, only the locking lug blades will exhibit this defect. Monitor this condition to see if a platform crack develops. Also look for missing pieces around the locking lugs. You must report and record any cracks in the platform in the following manner:

- The stage
- The number of blades

- The spacing of the blade numbers separating the shingled blade platforms
- The platform gap observation (Estimate gap as percent of circumferential span of the platform.)
- The condition of the shingled edge (bent, fretted, or stepped)

Mid-Span Shroud Wear. Some stage 1 compressor blades show wear at the mating 270.26 surfaces of the mid-span damper shrouds. Wearing of the tungsten-carbide wear coat causes the mating face contour to change from a straight line to a stepped line. This occurs at the after edge of the clockwise blade mid span (trailing edge) and the forward edge of the counterclockwise blade mid-span shroud (leading edge). In the step area, some metal may be turned or protruding from the mid-span shroud mating line (mushrooming). This protrusion is indicative of wear-through. A missing pad on one face would initiate an accelerated failure of the mating surfaces.

Blade Deposits. Compressor blades and stator vanes exhibit varying degrees of cleanliness. Variables such as air-inlet configuration, ambient atmospheric conditions, and air contaminants (chemicals, salt, dirt, water, and so forth) all tend to affect the surface condition of the compressor rotor and stator blades.

Aluminum Deposits. Two areas in the compressor assembly are coated with aluminum, the shrouding over the blade tips and the rotor drum area under the stator vanes. Tip rubs of either the blades or the vanes will rub off the aluminum coating. As time is accrued on the compressor assembly, the after stages of the rotor release the aluminum coating or flakes. This deterioration is a normal progression because of the differences in thermal expansion of dissimilar metals and the differences in the size and configuration of the various parts. The released aluminum flakes enter the airstream, impacting the rotor blades or vanes and splattering on the airfoils. Aluminum splatter observed forward of stage 11 can be caused by object damage and aluminum flakes that are rubbed out of the compressor case coating. This condition requires thorough inspection of the forward compressor stages.

Leading Edge Buildup. Aluminum buildup on the leading edges of blades is usually observed in stages 11 through 16. The buildup changes the contour of the airfoil and can affect the stall margin. You should report the presence of leading edge buildup in the inspection report. This type of buildup may occur on low-time compressors. The compressor blades tend to "self-clean" or lose this leading edge buildup as the assembly accrues time.

Airfoil Powdering. Compressor rotor blades may have aluminum particles visible on the airfoils in varying degrees (from stage to stage). This deposit is indicative of a possible compressor stall or a hard blade tip rub.

Combustion Section

Inspect the combustor for eroded or burned areas, cracks, nicks, dents, hot streaks, flatness of liners caused by hot spots, blocked air passages, and carbon buildup. If damage is found in the combustion section, it usually consists of a burn-through in the dome area adjacent to a fuel nozzle. The cause can usually be traced to a loss of film-cooling air due to upstream debris or to a faulty fuel nozzle. Cracking is not normally a problem, but you should photograph and report any suspected or confirmed cracks. Carbon deposits around the fuel nozzles occur on all engines and are not considered serious because the deposits build up only on the venturi and swirl cup rather than on the shroud or discharge orifice. They do not usually interfere with the fuel spray pattern. If you find cracking, evaluate it to ensure that no pieces will detach and cause any secondary damage to the HP turbine.

Combustion Section Damage. In the following paragraphs, we describe some of the damage that you might find during a borescope inspection of the combustion section. The dark surfaces in the combustion section will absorb light, necessitating the use of a 1,000-watt light source for proper inspection.

Discoloration. Normal aging of the combustor components will show a wide range of color changes, which should not be a cause for concern. As operating time is accrued on the combustor assembly, an axial streaking pattern running aft of every other circumferential fuel nozzle will occur. On low-time assemblies, the coloration is random and has little or no information to aid you during the inspection. As operating increases on the assembly, you will note significant deterioration at the edges of the streaking patterns. Cracking will begin in the forward inner liner panels and will propagate aft. The axial cracks tend to follow the light streaks. Panel overhang cracking and liberation usually occur at the edge of the streaks.

Riveted Joints. The dome band and the inner and outer liner assemblies are joined by rivets. The presence and condition of the rivet heads and rivet holes are easily assessed because of their position in relationship to the borescope ports. Record any missing rivets and torn or cracked hole edges.

Dome Assembly. Distortion of the trumpets and/or swirl cups is random and occurs on high-time assemblies. Record the distortion (in percent) of the edge and/or span of the trumpet and the percent of circumference versus diameter of the swirlers.

Cracking in the dimples of the dome bands occurs at relatively low operating time. Record the number of cracks and their relationship to one another. (Indicate if they are parallel, T-shaped, circumferential or angled to connect and separate part of the band, and so forth).

Record all the missing metal areas or burn-throughs. For the dome bands, estimate the magnitude by the number or partial/ circumferential span of the dimples and axially by percent of span of the band overhang to the trumpet. Record the trumpet areas of burn-away and burn through of the dome plate around the swirl cups. Burn-through in the combustor dome will reduce cooling flow to the HP turbine vanes. Monitor the HP turbine vane condition as burn-through progresses.

Igniter Tubes and Ferrules. Inspect the two igniter locations for the condition of the weld at the cutaway of the trumpet and the dome band. The ferrules are visible from these ports. Record the condition for evidence of cracking and/or loss of ferrule metal. Cracking from the igniter tube aft to the panel overhang is common.

Inner and Outer Liner Assemblies. You can inspect all areas of the inner and outer liner assemblies aft of the fuel nozzles by varying the immersion depth, rotating, and tilting the probe. Some of the damage that you may find is described in the following paragraphs.

Circumferential Cracks. High-time combustion liners show circumferential cracking, which occurs over the area of the inner liner stiffening bands. The bands are circumferential stiffeners and are not visible when viewed through the borescope inside the combustor assembly. Before actual cracking, the thermal working of the liner shows stress lines. These lines will be visible in all panels. Take care to inspect for the presence of cracks, not merely lines. A crack will be open and the separation will show an edge. The distortion occurs so that the inner liner lifts up into the flow path and the outer liner bends down into the flow path. These irregularities are usually obvious when the liners are viewed through the wide angle probe No. 2. When circumferential cracking is observed, record the band number and the span of the cracking relative to the number of cooling/dilution holes. Use the diameter of the cooling holes as a comparative measurement gauge.

Axial Cracks. Axial cracking usually starts at band No. 3 on the inner liners and propagates aft and forward. As operating time is accrued, these axial panel cracks grow into three-legged cracks. The edges of these cracks will separate and the corners will lift into the flow path. Inspect the areas aft and forward of these cracks, recording the axially separated cracks that show a tendency to grow together.

The primary cause of damage to the HP nozzle and turbine rotor elements is DOD. It is caused by pieces from the combustor liners cracking out of the panel overhangs and impacting with the rotating turbine elements. The most serious problem is the separation of a large selection of liner that could cause significant damage. This usually occurs as a result of axial and circumferential cracks growing together. It is important to record the damage to adjacent areas of about 5 inches to either side of the damaged area. These areas can grow together and liberate large pieces of material. These circumferentially cracked areas are usually separated at every other fuel nozzle spacing along with axial color streaking.

Missing Metal and Burn-Through. Inspect for the loss of metal at the panel overhang, the area between dimples. Burn through of the liners is not common although bluish-black slag areas will show roughness and appear to be oxidized. Inspect these areas carefully for T cracks because they will propagate and open up.

Distortion. Distortion or bowing of the liner assemblies is extremely difficult to assess when viewed through the borescope. If an axial streak (gutter) is observed to be out of contour, estimate the relative distortion in terms of dimples spanned or in relation to the diameter of the dilution holes. If the distortion is present at the number 1 band, estimate the contour change at the dome band relative to the panel.

HP Turbine

Inspect the HP turbine for eroded or burned areas, cracks or tears, nicks or dents, and missing blades. Knifing (erosion resulting in sharp edges) can occur on first-stage blades. The severity will vary according to the cleanliness of the turbine inlet air. Check for pitting on the leading edge near the root of the second-stage blading.

Cracking of the first-stage nozzle guide vanes is not too common, but photograph and report any suspected cracks. First-stage vane surfaces at the juncture of the inner and outer platforms have a tendency to corrode or erode. It would not be unusual for you to find several small penetrations during the service life. Most of these penetrations remain small and are not usually severe enough to warrant engine replacement. Record any such penetrations and regularly inspect them for change in size or quantity.

Vane HP (concave) surfaces will show gradual erosion with time, and the trailing edge slots will become elongated. When this degradation reaches maximum service limits, as noted on the PMS card or in the manufacturer's technical manual, the engine must be replaced.

HP turbine second-stage blades have a service life that is dependent upon operating conditions. Cracks are the major inspection criteria listed. You should document and report any confirmed cracks. The most common form of degradation is deposit buildup and erosion; this is not usually as severe as on the first-stage blades. The most serious form of damage that you may find in this area is pitting in the root area, which you must document and report.

HP TURBINE NOZZLE DAMAGE. The first-stage turbine nozzle vanes are inspected simultaneously with the combustor and fuel nozzles. In the following paragraphs, we describe the damage you may find during the borescope inspections.

Discoloration. Normal aging of the HP turbine nozzle stage 1 vanes will result in coloration changes as operating time is accrued. There is no limit relative to discoloration of HP turbine nozzle vanes.

Oxidation and/or burning of the vane areas are accompanied by dark areas silhouetting the initial distress. Cracks are shrouded in dark patches adjacent to the defect. Usually the distress starts as a crack, followed by oxidation of the shroud adjacent to the crack. Impact damage usually shows as a dark spot on the leading edge.

Leading Edge Damage. This type of damage can be found between the forward gill holes on the concave and convex side of the leading edge.

• Axial cracks form around the leading edge. Estimate the percent of span of the leading edge or span relative to the nose cooling hole rows to determine the crack length.

• Burns and spalling (leading edge) should not be construed as coloration only, but must have actual metal oxidized (surface metal loss), but no holes through the leading edge. Estimate the area boundaries by the nose cooling holes spanned both radially (up and down the leading edge) and axially (around or across the leading edge). Record the number of vanes affected.

• Blocked cooling air passages (leading edge) is another type of damage. If multiple hole blockage is observed, record the separation by the open cooling holes and the number of adjacent plugged holes.

Airfoil Concave Surface. Radial cracks run spanwise in the vane airfoil surface (up and down the vane). Record the relative chord position of the cracks. Record the relation of axial cracking versus radial cracking, such as axial and radial cracks that intersect or join at the second row of gill holes. The intent of the service limits are to preclude the liberation (break-out) of pressure face pieces.

Other Airfoil Area Defects. In the following paragraphs, we describe other defects that you may find during the inspections.

• Burns and cracks on concave and convex sides (charred). Record the area and length, estimate the length relative to the leading edge area (gill hole to gill hole and spanwise by span of cooling or gill holes). Estimate the surface damage relative to separation of gill hole rows and radially by gill or cooling holes.

• Craze cracking. These cracks are superficial surface cracks, caused by high temperature. They are random lines that are very thin in appearance with tight lines (no depth or width to the cracks). There is no limit against this condition.

• Nicks, scores, scratches, or dents. These defects are allowed by the service limit and may be present on any area of the nozzle vanes.

• Cracks in the airfoil fillet at the platform. There is no limit restricting these cracks, except at the leading edge area.

• Metal splatter. Aluminum and combustor liner metal, when liberated by the compressor or combustor, frequently splatter the surface areas of the stage 1 HP turbine nozzle vanes. There is no limit for these deposits; however, abnormal amounts of this splatter is reason to inspect the compressor.

Platforms. Cracking in the HP turbine nozzle stage 1 platforms is difficult to see from the combustor borescope ports. When this area is viewed through port No. 12, extreme magnification is afforded even with probe No. 2. This is due to the closeness of the surface to the distal end of the probe. Record the origin and end of the cracking and assess the magnitude using trailing edge slots and gill hole rows for radial and axial dimensions.

Nicks, scores, scratches, and dents on plat- form surfaces are again masked from the combustor ports, except for the forward areas.

Viewed via port No. 12, the area is magnified. Record the magnitude of the defect using the geometry of the trailing edge, gill hole rows, and gill hole separation for comparative dimensions.

You must record burns on vane platform areas and use probe No. 1 to assess the conditions. If a burn-through occurs, the inner and outer surface edge of the platform should be seen. This is a difficult assessment. Use of a fiberscope is recommended if the condition is doubtful. A confusing evaluation should be subject for a follow-up check condition after a specified operating time.

HP Turbine Blade Damage. When inspecting the HP turbine blades, you should use probe No. 2 with the 150-watt light source. In the following paragraphs, we describe some of the damage you may find.

Cracks in the Leading Edge. The leading edge of the stage 1 turbine rotor blades is the area forward of the gill holes. Cracks in the leading edge can be caused by DOD impact (combustion liner pieces) or thermal stress. An indication on the leading edge open enough to show depth is defined as a crack. Some conditions that will be observed that tend to mislead you in the determination of the presence of cracks are dirt and debris buildup in layers on the leading edge. When this buildup begins to flake off, the edge of the area where the flake came off causes visible lines. These lines are irregular and appear to be cracks. The other common point of confusion on leading edge cracks is on the convex side of the leading edge tip area. This area is subject to "scratching" by the small pieces of combustor metal that pass through the HP turbine.

Cracks in the Trailing Edge. The trailing edge is the flat surface with cooling holes and forms the after edge of the blade airfoil. Trailing edge cracks are difficult to see, but if a crack is suspected, use probe No. 1 for increased magnification. Record the location relative to a cooling hole and the magnitude of the crack. Record any plugged trailing edge cooling holes.

Cracks in Concave and Convex Surfaces. The airfoil surfaces are the areas aft of the gill holes back to the trailing edge. The tip area is further restricted to that area above the tip cap. When you evaluate the airfoil serviceability, do not consider the tip as a part of that area. Cracks in the airfoil surfaces are very tight, but can readily be seen with probe No. 2. They are irregular in edge appearance and are not misleading relative to streaks which are usually straight in appearance. Record the area by the percent of span or gill hole spacing equivalent for location and magnitude of the cracking. For axial position, use an estimate of percent chord and the position relative to the tip cooling film cooling holes.

Cooling Hole Blockage. The HP turbine rotor stage 1 blades are film cooled by air that flows out of the cooling holes. Report plugged holes relative to the number of blades affected and the position and number of plugged holes. Ensure the correct callout of the holes (such as the nose cooling, convex gill, tip film cooling holes, and so forth.)

Distortion. Heavy impact damage to the leading edge of the blade usually results in distortion. When the impact is severe enough, cracking and/or tearing of the leading edge, adjacent to the impact area, occurs. Record the magnitude and span location relative to the number of gill holes spanned. Estimate the out of contour as percent of width of the leading edge frontal area or relative to the lateral spanning of the leading edge cooling hole rows.

Blade Tip Nibbling. The HP turbine rotor stage 1 blade tip nibbling is associated with hot running engines. Momentary overtemperature operation (such as experienced during compressor stalls) has exhibited this type of deterioration. This area of the blade is above the tip cap and located about two-thirds of the chord aft from the leading edge.

Blade Leading Edge Impact Damage. The critical part of this type of damage is the axial or chord wire cracking which progresses from the impacted damaged area into the convex or concave airfoil surface.

Blade Coating Failure. The HP turbine protective coating is the key factor in the service life of an LM2500 GTE. The combined effect of film cooling and protective coating will extend the service life. Coatings are thinly and uniformly applied by a vacuum film deposition process and do not fail by chipping, peeling, or flaking. The normal failure modes are usually by pitting, rub off, or nicks and scratches. Occasionally a bubble will occur in the surface coating during the coating process. If a bubble occurs, it will be tested at the coating facility to ensure that it cannot be rubbed off the surface. These imperfections pose no problem to the engine. If the bubble area of the coating fails, you should monitor that area to determine any further deterioration. Development and testing of new coatings that are highly resistant to corrosion and erosion are in progress. The latest blade coating to be introduced in fleet engines are designated BC23. This replaces BC21 and extends the service life of these blades considerably.

HP Turbine Blade Failure Modes. Failures that you may observe during a borescope inspection include the following types:

• Corrosion of the coating. This appears as pitting of the coating primarily in the 80-percent span mid-chord region of the concave airfoil (thumbprint) side and the 20-percent span mid- chord region (rootprint). This corrosion/erosion has not been found on blades coated with BC23.

• Cracks in all areas of the blade, including radial cracks in the tips. Cracks generally start at the cooling holes.

• FOD/DOD, including nicks and dents.

• Aluminum spattering that appears as metallic deposits on the blade. This results from compressor tip rubs.

• HP turbine blade tip rubs. This results in coating removal and tip damage.

TURBINE MID-FRAME DAMAGE. In the following paragraphs, we describe damage that you may find when inspecting the turbine mid frame.

Discoloration. On low-time liners, the coloration is random and sometimes appears as a wavy surface. The coloration is random both axially and circumferentially. On high-time liners you may observe some axial carbon streaking. There are no service limits on discoloration.

Liner Cracking. Initial deterioration of the turbine mid-frame liners occurs at the forward inner liner flange in the form of axial cracking. It is difficult to determine the magnitude or length of a crack in this area. The area is immediately aft of the HP turbine stage 2 blade platforms. Small tight cracks will probably not be noticed. Of primary interest is that there are no cracks with visible turned up edges. If cracking is observed in the forward inner liner flange, you can use a fiberscope for a closer look to establish the extent of the crack and the adjacent area condition. Cracking can also occur around the leading edge weld beads on the strut fairings at both the inner and outer liner areas.

Liner Distortion. Turbine mid-frame distortion most commonly occurs in the 10 to 12 o'clock area of the outer liner forward flange. The only relative gauge available for comparative assessment (roundness/contour) is the HP turbine stage 2 blade tip arc and the stage 2 shroud contour. A fiberscope is recommended for the final assessment of any suggested distortion of the liner. A guide tube is generally needed to position the fiberscope.

Power Turbine

The most common problem in the power turbine section is usually a loss of the hardcoat on the tip shroud, resulting in notch wear and subsequent blade bending. This leads to fatigue failure of the airfoil. The actual loss of the hardcoat cannot be confirmed through the borescope. It can be confirmed by removal of the upper case and actual physical inspection of the tip shrouds. You can see the symptoms through the borescope by looking at the notch with probe No. 1. Uneven notch wear may indicate loss of the hardcoat. You should carefully inspect for any transverse cracks in the blade airfoil around the 10-percent span. Any cracking is cause for replacement of the power turbine. The power turbine first-stage blades also have a history of deposit buildup that leads to rotor unbalance and excessive vibration.

Cracks in Blades. Inspect the total airfoil, platform, and tip shrouds for evidence of cracks. If you suspect a specific area, use the high-magnification probe. You will see a limited amount of the stage 1 blading when viewing aft from the turbine mid-frame liner inspection ports. You can see more detail with a fiberscope or by viewing forward from the turbine exhaust duct. Cracks will show depth and under magnification will show edge material definition. Be sure to distinguish cracks from false indications such as smears and carbon streaks.

Nicks and Dents. Record these defects in relation to the percent span and percent chord for magnitude and location on the blade. Record also the condition of the blade material adjacent (at the extremities of the defect) to the observed defect. Record any cracking or sharpness of nicks or dents. Investigate smooth impact deformities to determine the origin of damage.

Wear. Inspect LP turbine rotor blade tip shroud interlocks or circumferential mating surface for wear at stage 1.

DIRT, COLORATION, PITTING, AND CORROSION. High-time LP turbine rotor assemblies may show airfoil surface irregularities that could be dirt accumulation, carbon buildup, pitting of the surfaces from particles in the gas stream, or corrosion of the blade material. Dirt and coloration are of little concern; however, pitting and corrosion may be significant.

Evaluating Physical Size

You determine physical size in several ways. By using dimensional data in the manufacturer's technical manual, you can estimate size by making a comparison to a known dimension in the field of view. Another way you can evaluate size, particularly in regard to cracks, is to use a lock wire of a known diameter. The lock wire is inserted into the field of view and placed next to the crack for size comparison. When using this method, you should ensure that the lock wire cannot fall inside the engine. Using an absolute reference for size, such as the lockwire or a known dimension, is more reliable than estimating by the appearance of size through the borescope.

Color Evaluations

During inspections, you can observe deposits or various forms of deterioration. This information will aid you in interpreting what you see by evaluating the color of the area or component. Color photographs taken through the borescope are an effective method to record the results of an inspection.

It is difficult to make accurate color interpretation. The only color indication that might give immediate cause for concern is aluminum oxide splatter in the hot section of the engine. Other colorations are normal and do not limit the service life of the engine.

Water Washing

Water washing is not currently listed on the MRCs as a prerequisite to borescope inspections; however, to get the best possible evaluation of the engine condition, water washing is recommended. Dirt and soft carbon deposits may obscure small cracks and pitting that could be missed if the engine were dirty.

Reporting Inspection Results

Unless a discrepancy is found, you do not have to enter routine borescope inspections in the Marine Gas Turbine Service Record (MGSTR); however, if the inspection was conducted as a troubleshooting procedure due to an engine malfunction or was ordered by a higher authority, you must log the inspection and note the findings. You must evaluate and report all major damage or exceeded service limits to NAVSEA. If the damage or wear is extensive, the engine must be replaced.

Troubleshooting

As a engineer, you will find that successful troubleshooting is a rewarding experience. Proper use of the manufacturer's technical manual will enhance your professional abilities and result in getting the job done right the first time. In this section we will discuss the use of the troubleshooting sections of the LM2500 GTE technical manual.

TROUBLESHOOTING TECHNIQUES

Troubleshooting is a systematic analysis of symptoms that indicate an equipment malfunction. These symptoms usually appear as deviations from the normal parameters. You must be able to recognize normal operating conditions to recognize abnormal operation. If you have a thorough knowledge of equipment systems and use logical reasoning, you will be able to solve most troubleshooting problems with little difficulty. The basic methods used during troubleshooting are as follows:

• Be sure that you know the normal operating conditions (be able to recognize a problem).

• Find out everything about the nature of the malfunction. Write down all the symptoms and see if they follow an identifiable pattern.

• Check the obvious.

1. Blown fuses
2. Tripped circuit breakers
3. Faulty alarms
4. Loose connectors and cannon plugs
5. Switches in the wrong position
6. Burned-out lamps
7. Physical damages
8. Last PMS or maintenance procedure performed
9. System alignment

LM2500 TROUBLE ISOLATION

The trouble isolation section (volume 2) of the LM2500 technical manual consists of three chapters with troubleshooting information that will aid you in isolating faults and malfunctions in the LM2500 GTE and its ancillary equipment. The manual presents troubleshooting procedures in fault logic diagrams, Functional Dependency Diagrams (FDDs), and signal flow diagrams

Fault Logic Diagrams

These diagrams are based on a fault indication observed during troubleshooting. The diagrams comprise a branching series of questions pertaining to fault isolation. Each question pertains to further observation or measurement, and results in a yes or no answer.

In this way, the possible functional area of the fault is progressively narrowed. Tolerance values are presented in those instances where a definitive yes or no is not obtained. This progression and elimination will isolate the functional area of the equipment containing the fault, and then refer you to the portion of the manual needed to complete the fault isolation and repair. Each diagram includes, or makes reference to, information necessary to establish the test or operating conditions required for starting the fault isolation procedure. Three types of blocks are used in the fault logic diagram.

- Shaded blocks (right and bottom border lines shaded). Contain questions that may be answered from observation, without changing the test setup and without special equipment

- Single-line blocks. Contain questions requiring measurement by special setup of external test equipment

- Double-line blocks (conclusion blocks). List the functional area within an equipment unit that is the probable source of the malfunction and reference a procedure or another diagram for further isolation or correction of a fault

Functional Dependency Diagrams

The FDD are used to support troubleshooting of the gas turbine electronic power control system. An FDD is a block diagram that illustrates the functional dependency of one test point (or circuit) upon another.

Signal Flow Diagram

The signal flow diagram depicts the circuitry for each of the main functions of the circuit that you are troubleshooting. The notes preceding the signal flow diagram contain instructions for establishing operating conditions and connecting test equipment that is required for measuring the circuit parameters.

THE ALLISON 501-K17/34 GTE

Maintenance and troubleshooting procedures for the Allison 501-K17/34 GTE are similar to the procedures used for the LM2500 engine. In all cases, you must use the proper EOSS, PMS, and technical manuals when conducting any maintenance or troubleshooting.

Volume 2 of the *Model 104/Model 139 Gas Turbine Generator Set* technical manual is divided into two parts. Part 1 contains all the necessary information, procedures, and diagrams for locating a malfunction. Part 2 is for corrective maintenance. It contains procedures for adjustment and alignment, repair, and removal/replacement of components.

When entering and working within the engine enclosure, follow the proper EOSS procedure and all standard safety precautions at all times.

Summary

In this chapter we have discussed object damage, borescope inspection, and troubleshooting related to GTEs. You have often been referred to the applicable technical manuals for specific information. You must use these manuals to guide you through the procedures. Proper use of the technical manuals will ensure that you make a complete inspection and/or properly isolate a problem.

CHAPTER 6

GAS TURBINE PRESERVATION AND CORROSION CONTROL

Modern gas turbines and their support equipment are dependent upon the structural soundness of the metals from which they are fabricated. The greatest threat to the structural integrity of this equipment is corrosion of the metal. With the higher demands being made on these metals, both in strength and in closer tolerances, this equipment would rapidly deteriorate and become inoperative without regular attention to corrosion control.

Corrosion endangers the gas turbine and its support equipment by reducing the strength and changing the mechanical characteristics of the materials used in their construction. All such materials are designed to carry certain loads and withstand given stresses and temperatures, as well as to provide an extra margin of strength for safety. Corrosion can weaken the structure, thereby reducing or eliminating this safety factor. Replacement or repair operations are costly, time consuming, and restrict the usage of the equipment. Corrosion in electronic and electrical components can cause serious malfunctions that reduce the effectiveness and reliability of the engineering plant and can often completely destroy these components.

A thorough comprehension of the dangers of corrosion and the ability to recognize and cope with the various types of corrosion should be included in the objectives of any maintenance training program. As a work center supervisor, you may find that corrosion prevention and control frequently turns out to be an all-hands evolution. To some extent this situation can be avoided through frequent inspections, the effective utilization of your available manpower, and the training of your subordinates.

Corrosion

The problem of protection of gas turbine engines and support equipment is threefold: (1) prevention of corrosion of the metal parts; (2) control of deterioration of non-metallic materials; and (3) elimination of physical damage during replacement, repair, and maintenance. Of the three, corrosion of metals is the most difficult to control.

Metal corrosion is the deterioration of a metal. When the metal is combined with oxygen it forms metallic oxides. This combining is a chemical process that is essentially the reverse of the process of smelting the metals from their ores. Very few metals occur in nature in the pure state. For the most part, they occur as metallic oxides. The refining process involves the extraction of relatively pure metal from its ore and the addition of other elements (both metallic and non-metallic) to form alloys.

After refining, regardless of whether or not they are alloyed, base metals possess a potential or tendency to return to their natural state. However, this potential is not enough in itself to initiate and promote this reversion. There must also exist a corrosive environment in which the significant element is oxygen. It is the process of oxidation that causes metals to corrode. It is a well-known fact that the tendency to corrode varies widely between various metals. For example, magnesium alloys are very difficult to protect, while copper alloys have relatively good corrosion resistance.

Corrosion may take place over the entire surface of a metal by a chemical reaction with the surrounding environment, or it may be electro-chemical in nature between two different metallic materials or two points on the surface of the same alloy that differ in chemical activity. The presence of some type of moisture is usually essential in both types of attacks.

Causes

Prevention and control of corrosion begins with an understanding of the causes and nature of this phenomenon. As stated earlier, corrosion is caused by an electro-chemical or a direct chemical reaction of a metal with other elements. In the direct chemical attack, the reaction is similar to that which occurs when acid is applied to bare metal. Corrosion in its most familiar form is a reaction between metal and water and is electro-chemical in nature.

In the electro-chemical attack, metals of different electrical potential are involved and they need not be in direct contact. When one metal contains positively charged ions and the other contains negatively charged ions and an electrical conductor is bridged between them, current will flow as in the discharge of a dry cell battery. In this type of reaction, the conductor bridge may be any foreign material such as water, dirt, grease, or any debris that is capable of acting as an electrolyte. The presence of salt in any of the foregoing media tends to accelerate the current flow and hence speed the rate of corrosive attack.

Once the electrolyte has completed the circuit, the electron flow is established within the metal in the direction of the negatively charged area (cathode), and the positively charged area (anode) is eventually destroyed. All preventive measures taken with respect to corrosion prevention and control are designed primarily to avoid the establishment of the electrical circuit, or secondly, to remove it as soon as possible after its establishment before serious damage can result.

Electro-chemical attack is evidenced in several forms depending on the metal involved, its size and shape, its specific function, the atmospheric conditions, and the type of corrosion producing agent (electrolyte) present. A great deal is known about the many forms of metal deterioration that results from electro-chemical attack. But despite extensive research and experimentation, there is still much to be learned especially about other more complex and subtle forms. Since there are so many factors which contribute to the process of corrosion, selection of materials by the manufacturer must be made with weight versus strength as a primary consideration and corrosion properties as a secondary consideration. However, close attention during design and production is given to heat treating and annealing procedures, protective coatings, choice and application of moisture barrier materials, dissimilar metal contacts and accesses. Every logical precaution is taken by manufacturers to inhibit the onset and spread of corrosive attack.

There are many factors that affect the type, speed, cause, and seriousness of metal corrosion. Some of these factors can be controlled; others cannot. Preventive maintenance such as inspection, cleaning, painting and preservation are within the control of engineering personnel. Preventive maintenance offers the most positive means of corrosion deterrence.

The electro-chemical reaction which causes metal to corrode is a much more serious factor under wet, humid conditions. The salt in seawater and in the air is the greatest single cause of corrosion. Hot environments speed the corrosion process because the electro-chemical reaction develops faster in a warm solution, and warm moisture in the air is usually sufficient to start corrosion if the metal surfaces are unprotected.

Another corrosion factor is the relationship between dissimilar metals. When two dissimilar metals come in contact and the more active metal is small compared to the less active one, corrosive attack will be severe and extensive. Insulation between such contact will inhibit this process. However, if the area of the less active metal is small compared to the other metal, corrosive attack will be relatively slight.

CHARACTERISTICS

The appearance of corrosion will vary with the metal involved. The following discussion includes brief descriptions of typical corrosion product characteristics of the most common materials used in gas turbine propulsion and support equipment.

Iron and Steel

Possibly the best known and easiest recognized of all forms of metal corrosion is the familiar reddish-colored iron rust. When iron and its alloys corrode, dark iron oxide coatings usually form first. These coatings, such as heat scale on steel sheet stock and the magnetite layer that forms on the inside of boiler tubes, protect iron surfaces rather efficiently. However, if sufficient oxygen and moisture are present, the iron oxide is soon converted to hydrated ferric oxide, which is conventional red rust.

Aluminum

Aluminum and its alloys exhibit a wide range of corrosive attacks, varying from general etching of aluminum surfaces to penetrating attacks along the internal grain boundaries of the metal. The corrosion products of aluminum are seen as white gray powdery deposits.

Copper and Copper Alloys

Copper and its alloys are generally corrosion resistant, although the products of corrosive attack on copper are commonly known. Sometimes copper or copper alloy surfaces will tarnish to a gray-green color, while the surface will remain relatively smooth. This discoloration is the result of the formation of a fine-grained, airtight copper oxide crust, called a patina.

This patina in itself offers good protection for the underlying metal in ordinary situations. However, exposure of copper alloys to moisture or salt spray will cause the formation of blue or green salts called verdigris, indicating active corrosion. These salts will form over the patina which is not impervious to water.

Cadmium and Zinc

Cadmium is used as a coating to protect the area to which it is applied and to provide a compatible surface when the part is in contact with other metals. The cadmium plate supplies sacrificial protection to the underlying metal because of its great activity. During the time it is protecting the base metal, the cadmium is intentionally being consumed. Zinc coatings are used for the same purpose, although to a lesser extent. Attack is evident by white-to-brown-to-black mottling of the surfaces. These indications do not indicate deterioration of the base metal. Until the characteristic colors peculiar to corrosion of the base metal appear, the coating is still performing its protective function.

Nickel and Chromium Alloys

Nickel and chromium alloys are also used as protective agents, both in the form of electroplated coatings and as alloying constituents with iron in stainless steels, and with other metals such as copper. Nickel and chromium plate provide protection by the formation of an actual physical noncorrosive barrier over the steel. Electroplated coatings, particularly chromium on steel, are somewhat porous, and corrosion eventually starts at these pores unless a supplementary coating is applied and maintained.

TYPES OF CORROSION

As stated previously, corrosion may occur in several forms depending upon the metal involved, its size and shape, its specific function, the atmospheric conditions, and the corrosion-producing agents present.

Those described in this section are the most common forms found on gas turbine engines and machinery structures.

Direct Surface Attack

The surface effect produced by reaction of the metal surface to oxygen in the air is a uniform etching of the metal. The rusting of steel, tarnishing of copper alloys, and the general dulling of aluminum surfaces are common examples of a direct surface attack. If such corrosion is allowed to continue unabated, the surface becomes rough, and in the case of aluminum, frosty in appearance.

Galvanic Corrosion

Galvanic corrosion is the term applied to the accelerated corrosion of metal caused by dissimilar metals being in contact in a corrosive medium. Dissimilar metal corrosion is usually a result of faulty design or improper maintenance practices which result in dissimilar metals coming in contact with each other. This is usually seen as a buildup of corrosion at the joint between the metals. For example, when aluminum pieces are attached together with steel bolts and moisture or contamination is present, galvanic corrosion occurs around the fasteners.

Pitting

The most common effect of corrosion on aluminum alloys is called pitting. It is due primarily to variations in the grain structure between adjacent areas on the metal surface that is in contact with a corrosive environment. Pitting is first noticeable as a white or gray powdery deposit, similar to dust that blotches the surface. When the superficial deposit is cleaned away, tiny pits or holes can be seen in the surface. They may appear either as relatively shallow indentations or as deeper cavities of small diameters. Pitting may occur in any metal, but it is particularly characteristic of aluminum and aluminum alloys.

Intergranular Corrosion

Intergranular corrosion is an attack on the grain boundaries of some alloys under specific conditions. During heat treatment, these alloys are heated to a temperature which dissolves the alloying elements. As the metal cools, these elements combine to form other compounds. If the cooling rate is slow, they form predominantly at the grain boundaries. These compounds differ electrochemically from the metal adjacent to the grain boundaries and can be either anodic or cathodic to the adjoining areas, depending on their composition. The presence of an electrolyte will result in an attack on the anodic area. This attack will generally be quite rapid and can exist without visible evidence.

As the corrosion advances, it reveals itself by lifting up the surface grain of the metal by the force of expanding corrosion products occurring at the grain boundaries just below the surface. This advanced attack is referred to as exfoliation. Recognition and necessary corrective action to immediately correct such serious corrosion are vital. This type of attack can seriously weaken structural members before the volume of corrosion products accumulate on the surface and the damage becomes apparent.

Fretting

Fretting is a limited but highly damaging type of corrosion caused by a slight vibration, friction, or slippage between two contacting surfaces which are under stress and heavily loaded. It is usually associated with machined parts such as the area of contact of bearing surfaces, two mating surfaces, and bolted assemblies. At least one of the surfaces must be metal.

In fretting, the slipping movement at the interface of the contacting surface destroys the continuity of the protective films that may be present on the surfaces. This action removes fine particles of the basic metal. The particles oxidize and form abrasive materials which further accumulate and agitate within a confined area to produce deep pits. Such pits are usually located where they can increase the fatigue potential of the metal.

Fretting is evidenced at an early stage by surface discoloration and by the presence of corrosion products in any lubrication present. Lubricating and securing the parts so that they are rigid are the most effective measures for the prevention of this type of corrosion.

Stress

Stress, evidenced by cracking, is caused by the simultaneous effects of tensile stress and corrosion. Stress may be internal or applied. Internal stresses are produced by nonuniform deformation during cold working, by unequal cooling from high temperatures during heat treatment, and by internal-structural rearrangement involving volume changes. Stresses set up when a piece is deformed, those induced by press-and-shrink fits, and those in rivets and bolts are examples of internal stresses.

Concealed stress is more important than design stress, especially because stress corrosion is difficult to recognize before it has overcome the design safety factor. The magnitude of the stress varies from point-to-point within the metal. Stresses in the neighborhood of the yield strength are generally necessary to promote stress corrosion cracking, but failures may occur at lower stresses.

Fatigue

Fatigue is a special type of stress corrosion. It is caused by the combined effects of corrosion and stresses applied in cycles to a part. An example of cyclic stress is the alternating loads to which the connecting rod of a double-acting piston in an air compressor is subjected. During the extension (up) stroke a compression load is applied, and during the retraction (down) stroke a tensile or stretching load is applied. Damage from fatigue is greater than the combined damage of corrosion and cyclic stresses than if the part was exposed to each separately. Fracture of a metal part due to fatigue corrosion generally occurs at a stress far below the fatigue limit in a laboratory environment, even though the amount of corrosion is very small. For this reason, protection of all parts subject to alternating stress is particularly important wherever practical, even in environments that are only mildly corrosive.

PREVENTION AND CONTROL

Much has been done over the years to improve the corrosion resistance of newer Navy warships and combatant crafts, including improvements in the selection and combination of materials of construction, in chemical surface treatments, in insulation of dissimilar metals, and in protective paint finishes. All of these are aimed at reducing maintenance as well as improving reliability. But in spite of refinements in design and construction, corrosion control is a problem that requires a continuous maintenance program, considering the adverse environment of a ship and craft at-sea.

Cleaning

As a professional engineer, one of your most important aids in the prevention and control of corrosion is an adequate cleaning program. The term clean means to do the best job possible using the time, materials, and personnel available. A daily wipe down of all machinery is better than no cleaning at all. The importance of frequent cleaning cannot be overemphasized. Any cleaning procedures, however, should be in the mildest form possible to produce the desired results. For example, spraying water around multi-pin connectors can possibly cause electrical shorts or grounds, with a possible loss of control functions or damage to equipment.

In general, gas turbine engines and enclosures should be cleaned as often as necessary to keep surfaces free of salt, dirt, oil, and other corrosive deposits. A thorough inspection and cleaning of gas turbine intakes and enclosures should always be accomplished prior to getting underway after an extended stay in port, and after returning to port from an extended time at sea, as well as in conformance with PMS requirements.

Since marine gas turbines are more subject to internal corrosion than engines used in other types of applications, internal cleaning is of particular importance. This is accomplished by means of water washing. A mixture of B & B 3100 water wash compound and distilled water is injected into the engine air inlet while it is being motored and then rinsed with distilled water in the same manner. It is then operated for about 5 minutes to remove all liquid. For more detailed information on this procedure, consult the applicable PMS card for the engine to be water washed.

Characteristics of Metals

All engineers are expected to have a good knowledge of the characteristics of the various metals used throughout the engineering plant, as well as the engines themselves. To some extent, all metals are subject to corrosion. In order to keep corrosion to a minimum, corrosion-resistant metals are used to the fullest extent possible consistent with weight, strength, and cost consideration. On exposed surfaces, the principal preventative relied upon to provide relative freedom from corrosion is a coating of a protective surface film in the form of an electroplate, paint, or chemical treatment, whichever is most practical.

Most of the metals used in the engineering plants require special preventive measures to guard against corrosion. In the case of aluminum alloys, the metal is usually anodized or chemically treated and painted. Steel and other metals such as brass or bronze (with the exception of stainless steels) use cadmium or zinc plating, protective paint, or both. In all cases, the protective finish must be maintained in order to keep active corrosion to an absolute minimum.

Preservation and Depreservation of Gas Turbine Engines

The purpose of engine preservation is to prevent corrosion of the various types of materials that make up the engine and its accessories, and to ensure against gumming, sticking, and corrosion of the internal passages.

Engine preservation and de-preservation is of vital concern to engineers because the corrosion of engine structures can and does have a great effect on the operational and structural integrity of the unit. Therefore, it is important that engineers have a knowledge of the method of preservation, materials used, and de-preservation procedures.

Preservation and Packaging for Storage

If it is known that an engine is to be shipped or stored, it must be preserved prior to removal from the ship. Engines to be taken out of operation for periods of up to 1 month require only that the unit be protected from the elements, but units that will be stored or out of service for more than a month must be preserved for storage.

Packaging for storage should be in compliance with current instructions for engine shipment furnished by the manufacturers. If specific manufacturer's instructions are not available, then the engine should be placed in a hermetically sealed metal container with a humidity control and an external humidity indicator.

All major engine parts, no matter how badly worn or damaged, must be returned with the engine whether it is to be overhauled or salvaged; and the entire assembly (engine and accessories) must be protected from damage during shipment. When preparing the engine for shipment, you must ensure that all fuel lines, receptacles, oil lines, intakes, exhausts, and any other openings in the engine or its components are capped or covered before the engine is removed.

Depreservation

An engine that has been in storage, or inoperable for an extended period of time, must be depreserved before it can be placed in service. Prior to connecting the engine to the external portion of the fuel and oil system (supply tank, coolers, filters, and so forth), the external tubing and equipment must be thoroughly flushed and purged. After installation, fill the oil sump (Allison) or LOSCA (LM2500) with clean lubricating oil to the proper operating level.

Caution

To prevent accidental firing, ensure that the engine ignition circuit is disconnected when priming the fuel control and the fuel system.

Before initial operation, the engine fuel system must be flushed and purged. To accomplish this, the engine is motored until all bubbles are out of the fuel stream and only fuel comes through. While motoring, observe the engine oil pressure. If no pressure is indicated, the cause must be determined and corrected before the engine can be started. In all cases, the manufacturers' technical manual must be consulted for specific instructions on the de-preservation and startup of each particular engine.

Summary

In this chapter we have discussed corrosion, its causes, effects, and some of the methods available to us to combat and minimize it. Since this chapter is not written to address all aspects of any one subject, you are urged to study the various publications on the prevention and control of corrosion mentioned in this chapter. Of particular interest are the Propulsion Gas Turbine Manual LM2500, volume 2, part 2, S9234-AD- MMO-0401 LM2500, the Naval Ships' Technical Manual, chapter 234, "Marine Gas Turbine," Corrosion Control and Prevention Manual, NAVSEA S9630-AB-MAN-010; and the Standard Corrosion Control Manual, NAVSEA S9630-AE-MAN-010.

REFERENCES

1. Introduction to Marine Gas Turbines, NAVEDTRA 10094, Naval Education and Training Program Development Center, Pensacola, Fla., 1978

2. S9234-AL-GTP-010/00963 PPM

3. Propulsion Gas Turbine Manual LM2500, volume 2, part 2, S9234-AD-MMO-0401 LM2500

4. Naval Ships Technical Manual, chapter 234, "Marine Gas Turbine,"

5. Corrosion Control and Prevention Manual, NAVSEA S9630-AB-MAN-010

6. Standard Corrosion Control Manual, NAVSEA S9630-AE-MAN-010.

APPENDIX I

GLOSSARY

AMPERE (amp). A unit of electrical current or rate of flow of electrons. One volt across one ohm of resistance causes a current flow of one ampere.

ANALOG SIGNAL. A measurable quantity that is continuously variable throughout a given range and is representative of a physical quantity.

ANNULAR. In the form of, or forming a ring.

ANTI-ICING. A system for preventing buildup of ice on the gas turbine intake systems.

AUTOMATIC PARALLELING DEVICE (APD). Automatically parallels any two generators when initiated by the EPCC.

BABBITT. A white alloy of tin, lead, copper, and antimony that is used for lining bearings.

BLEED AIR. Hot compressed air bled off the compressor stages of the GTMs and GTGS.

See BLEED AIR SYSTEM.

BLEED AIR SYSTEM. This system uses as its source compressed air extracted from the compressor stage of each GTM or GTGS. It is used for anti-icing, prairie air, masker air, and LP gas turbine starting for both the GTMs and GTGS.

BLOW-IN DOOR. Doors located on the high hat assembly designed to open by solenoid-operated latch mechanisms if the inlet air flow becomes too restricted for normal engine operation.

BORESCOPE. A small periscope (instrument) used to visually inspect internal engine components.

BUFFER. Used to electronically isolate and filter an electrical signal from its source.

BUS. An uninsulated power conductor (a bar of wire).

CIRCUIT BREAKER (CB). An automatic protective device that, under abnormal conditions, will open a current-carrying circuit.

CLUTCH/BRAKE ASSEMBLY. A clutch/brake assembly for each GTM engine is mounted on the MRG housing to couple or decouple either or both engines to the drive train, to stop and hold the power turbine, and for shaft braking.

COMPRESSOR DISCHARGE PRESSURE (CDP). Compressor discharge pressure is sensed by a pressure tap on the compressor discharge static pressure sensing line to the MFC and piped to a base-mounted transducer on the GTM.

COMPRESSOR INLET TEMPERATURE (CIT OR T2). The temperature of the air entering the gas turbine compressor (GTM) as measured at the front frame; one of the parameters used for calculating engine power output (torque) and scheduling combustion fuel flow and variable stator vane angle.

COMPRESSOR INLET TOTAL PRESSURE (Pt2). Sensed by a total pressure probe mounted in the GTM compressor front frame.

CONTROL AIR SYSTEM. The equipment that controls the compressed air used to operate the main clutch/brake assemblies. One control air system unit is mounted on each side of each MRG.

CONTROLLABLE REVERSIBLE PITCH (CRP) PROPELLER. A propeller whose blade pitch can be varied to control amount of thrust in both ahead and astern directions.

DAMAGE CONTROL CONSOLE (DCC). Located in CCS and provides monitoring for hazardous (fire, high bilge levels, and so forth) conditions. It also monitors the ship's firemain and can control the fire pumps.

DEAERATOR. A device that removes air from oil as in the lube oil storage and conditioning assembly tank (GTM) that separates air from the scavenge oil.

DEAERATING FEED TANK. A device used in the waste heat boiler system to remove dissolved oxygen and non-condensible gasses from the feedwater.

DIGITAL DISPLAY INDEX (DDI). A numerical display at the consoles that is used to read values of parameters within the engineering plant.

DEMISTERS. A moisture removal device (GTM intake system) that separates water from air.

DIFFUSER. A device that reduces the velocity and increases the static pressure of a fluid passing through a system.

EDUCTOR. A mixing tube (jet pump) that is used in the GTM exhaust system. It is physically positioned at the top of the stack so the gas flow from the GTM exhaust nozzles will draw outside air into the exhaust steam as it enters the mixing tube. It may also be a liquid pump used to dewater bilges and tanks.

ELECTRIC PLANT CONTROL CONSOLE (EPCC). Contains the controls and indicators used to remotely operate and monitor the generators and the electrical distribution system.

ELECTRIC PLANT CONTROL ENCLOSURE (EPCE). Provides centralized remote control of the GTGS and electrical distribution equipment. The EPCE includes the EPCC and is located in CCS.

ELECTRONIC GOVERNOR (EG). A system that uses an electronic control unit with an electrohydraulic governor actuator to control the position of the LFV on the GTGS and regulate engine speed.

ENGINEERING CONTROL AND SURVEILLANCE SYSTEM (ECSS). An automatic electronic control and monitoring system using analog and digital circuitry to control the propulsion and electric plant. The ECSS consists of the PLOE, PAMCE, SHIP CONTROL EQUIPMENT, EPCE, and PAMISE.

ENGINEERING OPERATIONAL SEQUENCING SYSTEM (EOSS). A two part system of operating instructions bound in books for each watch station. It provides detailed operating procedures (EOP) and casualty control procedures (EOCC) for the propulsion plant.

FAULT ALARM. This type of alarm is used in the FO control system and the DCC. It indicates that a sensor circuit has opened.

FOREIGN OBJECT DAMAGE (FOD). Damage as a result of entry of foreign objects into a gas turbine inlet.

FREE STANDING ELECTRONIC ENCLOSURE (FSEE). Provides the supporting electronics and engine control interface between the GTM and the control consoles. One FSEE is located in each MER.

FUEL OIL (FO) SYSTEM. Provides a continuous supply of clean fuel to the GTMs.

FUEL SYSTEM CONTROL CONSOLE (FSCC). Located in CCS and is the central station for monitoring and control of the fuel fill and transfer system (DDG-993, and CG-47 classes).

FULL POWER CONFIGURATION. The condition in which both engines (GTM) of a set are engaged and driving the reduction gear/ propeller shaft.

GAS GENERATOR (GG). The HP section of the main propulsion GT. It includes the compressor, combustor, HP turbine, front frame, compressor rear frame, turbine mid frame, transfer gearbox, and the controls and accessories.

GAS GENERATOR SPEED (Ngg). Sensed by a magnetic pickup on the aft transfer gearbox of the GTM.

GAS TURBINE (GT). Gas turbine engines provide main propulsion power and drive the SS generator sets.

GAS TURBINE GENERATOR SET (GTGS). Consist of a reduction gearbox, and a three-phase alternating current generator rated at 2000 or 2500 kw and 450 volts a.c.

GAS TURBINE MODULE (GTM). Consists of the main propulsion gas turbine unit including the GTE, base, enclosure, shock mounting system, fire detection and extinguishing system, and the enclosure environmental control components.

GENERATOR CONTROL UNIT (GCU). A static GCU is supplied for each GTGS consisting of a static exciter/voltage regulator assembly, field rectifier assembly, motor-driven rheostat, and a mode select rotary switch. It controls the output voltage of the generator.

GOVERNOR DROOP MODE. Droop mode is normally used only for paralleling with shore power. Since shore power is an infinite bus (fixed frequency), droop mode is necessary to control the load carried by the generator. If a generator paralleled with shore power and one attempts to operate in isochronous mode instead of droop mode, the generator governor speed reference can never be satisfied because the generator frequency is being held constant by the infinite bus.

If the generator governor speed reference is above the shore power frequency, the load carried by the generator will increase beyond capacity (overload) in an effort to raise the shore power frequency. If the speed reference is below the shore power frequency, the load will decrease and reverse (reverse power) in an effort to lower the shore power frequency. The resulting overload or reverse power will trip the GB.

GOVERNOR ISOCHRONOUS MODE. Normally used for generator operation. This mode provides a constant frequency for all load conditions. When operating two (or more) generators in parallel, isochronous mode also provides equal load sharing between units.

HEADER. A piping manifold that connects several sublines to a major pipeline.

HELIX. A tube or solid material (gear teeth) wrapped like threads on a screw.

HERTZ (Hz). A unit of frequency equal to one cycle per second.

HIGH HAT ASSEMBLY. A removable housing over the main engine air intake ducts that contains the moisture separation system (demisters), inlet louvers, and blow-in doors.

HIGH-PRESSURE (HP) AIR (3000 psig) SYSTEM. Used for emergency starting of the GTMs and GTGS. It is also used in operating the ASROC launcher, 5-inch guns, torpedo tubes, helicopter services, and as a backup emergency supply for the SSAS.

HUMIDITY, ABSOLUTE. The weight of water vapor in grains per cubic foot of air.

HUMIDITY, RELATIVE. The ratio of the weight of water vapor in a sample of air to the weight of water vapor that same sample of air contains when saturated.

HUMIDITY, SPECIFIC. The weight of water vapor in grains per pound of dry air.

INFRARED (IR) SUPPRESSION. Used to reduce GTGS exhaust temperature by injecting a seawater spray into the exhaust gases. This is designed to reduce the possibility of detection by heat-seeking devices.

INLET GUIDE VANES (IGVs). Vanes ahead of the first stage of compressor blades of a GTE. Their function is to guide the inlet air into the GT compressor at the optimum angle.

INLET PLENUM. That section of the GTE inlet air passage that is contained within the engine enclosure. Applies to GTM and GTGS engines.

LABYRINTH/HONEYCOMB SEALS. Combine a rotating element and a honeycomb stationary element to form an air seal. Used in the GTM turbine to maintain close tolerances over a large temperature range.

LABYRINTH/WINDBACK SEALS. Combine a rotating element with a smooth surface stationary element to form an oil seal. This type of seal is used with an air seal with a pressurization air cavity between the two seals. Pressure in the pressurization air cavity is always greater than the sump pressure. Therefore, flow across the seal is toward the sump. This prevents oil leakage from the sump. The windback is a coarse thread on the rotating element of the oil seal which uses screw action (windback) to force any oil that might leak across the seal back into the sump.

LIQUID FUEL VALVE (LFV). Meters the required amount of fuel for all engine operating conditions for the GTGS engine.

LOCAL CONTROL. Start-up and operation of equipment by manual controls attached to the machinery or by electric panel attached to the machinery or located nearby.

LUBE OIL STORAGE AND CONDITIONING ASSEMBLY (LOSCA). Mounted remotely from the GTM and is a unit with a lube oil storage tank, heat exchanger, scavenge oil duplex filter, and scavenge oil check valve (all mounted on a common base). Its function is to provide the GTM with an adequate supply of cooled, clean lube oil. It also has instrumentation for remote monitoring of oil temperature, filter differential pressure, and high/low tank level alarm.

MAIN FUEL CONTROL (MFC). A hydro-mechanical device on the propulsion gas turbine that controls N_{GG} schedules acceleration fuel flow, deceleration fuel flow, and stator vane angle for stall-free, optimum performance over the operating range of the GT.

MAIN REDUCTION GEAR (MRG) ASSEMBLY. A locked train, double reduction gear used on gas turbine ships. It allows the PT and the CRP propeller to operate at the most efficient speed.

MASKER AIR SYSTEM. Disguises the signature of the ship's hull and alters transmission of machinery noise to the water by emitting air from small holes in the emitter rings on the ship's hull. This reduces the reliability of ship identification by sonar.

MICRON. A unit of length equal to one millionth of a meter.

MIL. A unit of length equal to one thousandth of an inch.

NAVY DISTILLATE. A Navy classification of a distillate FO.

NOZZLE. A small jet (hole) at the end of a pipe.

ORIFICE. A restricted opening used primarily in fluid systems.

PERMANENT MAGNET ALTERNATOR (PMA). Mounted on the generator shaft extension of each GTGS and supplies speed sensing and power to the EG. The PMA also supplies initial generator excitation.

PINION. A smaller gear designed to mesh with a larger gear.

PITCH. A term applied to the distance a propeller will advance during one revolution.

POPPET-TYPE CHECK VALVE. A valve that moves into and from its seat to prevent oil from draining into the GTGS when the engine is shut down.

POUNDS PER SQUARE INCH (psi). Unit of pressure.

POUNDS PER SQUARE INCH ABSOLUTE (psia). Unit of pressure.

POUNDS PER SQUARE INCH DIFFERENTIAL (psid). Unit of pressure. Also known as delta (A) pressure.

POUNDS PER SQUARE INCH GAUGE (psig). Unit of pressure.

POWER LEVER ANGLE (PLA). A rotary actuator mounted on the side of the GTM fuel pump and its output shaft lever. It is mechanically connected to the MFC power lever. The PLA actuator supplies the torque to position the MFC power lever at the commanded rate.

POWER TAKEOFF (PTO). The drive shaft between the GTGS gas turbine engine and the reduction gear. Transfers power from the gas turbine to the reduction gear to drive the generator.

POWER TURBINE (PT). The GTM turbine that converts the GG exhaust into energy and transmits the resulting rotational force via the attached output shaft.

POWER TURBINE INLET TEMPERATURE (T5.4). Temperature sensed by thermocouples installed in the GTM mid frame.

POWER TURBINE TOTAL INLET PRESSURE (Pt5.4). Pressure sensed by five total pressure probes located in the GTM turbine mid frame and piped to a transducer on the bottom of the GTM.

POWER TURBINE SPEED (NPf). GTM power turbine speed is sensed by magnetic pickups in the GTM turbine rear frame.

PRAIRIE AIR SYSTEM. Disguises the sonar signature of the ship's propellers by emitting cooled bleed air from small holes along the leading edges of the propeller blades. The resulting air bubbles disturb the thrashing sound so identification of the type of ship through sonar detection becomes unreliable.

PRINTED CIRCUIT BOARD (PCB). An electronic assembly mounted on a card, using etched conductors. Also called printed wire board.

PROPULSION AND AUXILIARY CONTROL CONSOLE (PACC). Located in CCS and part of the PAMCE. It contains the electronic equipment capable of controlling and monitoring both propulsion plants and auxiliary equipment.

PROPULSION AND AUXILIARY MACHINERY CONTROL EQUIPMENT (PAMCE). Located in CCS and is part of the ECSS and includes the PACC and PACEE. This equipment provides centralized control and monitoring of both main propulsion plants and auxiliary machinery.

PROPULSION CONTROL CONSOLE (PCC). It is part of the PLOE. It has controls and indicators necessary for operator's control of one main propulsion plant and its supporting auxiliaries.

PROPULSION LOCAL OPERATING EQUIPMENT (PLOE). Located in each engine room and is part of the ECSS. It includes the PLCC and PLCEE. The PLOE provides for local control and monitoring of the main propulsion GTE and the associated auxiliary equipment.

RESISTANCE TEMPERATURE DETECTOR (RTD). These temperature sensors work on the principle that as temperature increases, the conductive material exposed to this temperature increases its electrical resistance.

SALINITY INDICATOR. An indicator used for measuring the amount of salt in a solution.

SCAVENGE PUMP. Used to remove oil from a sump and return it to the oil supply tank.

SECURED PLANT CONFIGURATION (MODE). The condition in which engines of a set (GTM) are disengaged from the reduction gear/propulsion shaft.

SENSOR. A device that responds to a physical stimulus and transmits a result impulse for remote motoring.

SHIP'S SERVICE AIR SYSTEM (SSAS). Supplies LP compressed air at 150 psig and 100 psig to a majority of the ship's pneumatically operated equipment.

SPLIT PLANT CONFIGURATION (MODE). The condition in which only one engine of a set (GTM, A or B) is engaged and driving the reduction gear/propulsion shaft.

STALL. An inherent characteristic of all gas turbine compressors to varying degrees and under certain operating conditions. It occurs whenever the relationship between air pressure, velocity, and compressor rotational speed is altered to such an extent that the effective angle of attack of the compressor blades becomes excessive, causing the blades to stall in much the same manner as an aircraft wing.

SUMMARY ALARM/FAULT. An indicator at a console that indicates to an operator that one of several abnormal conditions has occurred on a certain piece of equipment.

SWITCHBOARD (SWBD). A large panel assembly that mounts the control switches, CBs, instruments, and fuses essential to the operation and protection of electrical distribution systems.

TRANSDUCER. A sensor that converts quantities such as pressure, temperature, and flow rate into electrical signals.

TURBINE INLET TEMPERATURE (TIT). The GTGs turbine inlet temperature.

TURBINE OVERTEMPERATURE PROTECTION SYSTEM (TOPS). A system found on DDG and CG class ships used to protect a surviving generator from overload in the event of another generator failure.

ULTRAVIOLET (UV) FLAME DETECTORS. Sense the presence of fire in the GTM and GTGS and generate an electrical signal that is set to the ECSS.

VARIABLE STATOR VANE (VSV). Compressor stator vanes that are mechanically varied to provide optimum, stall-free compressor performance over a wide operating range. The inlet guide vanes (IGVs) and stage 1 through 6 stator vanes of the main propulsion gas turbine compressors are variable.

WASTE HEAT BOILER. Each waste heat boiler is associated with a GTGS and uses the hot GT exhaust to convert feedwater to steam for various ship's services.

APPENDIX II

ABBREVIATIONS AND ACRONYMS

This appendix is a listing of the abbreviations and acronyms used in this text. Although this is an extensive listing, it is not an all-inclusive list of abbreviations and acronyms used.

A

a.c.—alternating current
ACC—auxiliary control console
ACO—action cutout
APD—automatic paralleling device

C

CB—circuit breaker
CCS—central control station
CDP—compressor discharge pressure
CIT—compressor inlet temperature
CO2—carbon dioxide
CODAG—combined diesel and gas
CODOG—combined diesel or gas
COGOG—combined gas or gas
COSAG—combined steam and gas
CRP—controllable reversible pitch

D

d.c.—direct current
DCC—damage control console
DDI—demand display indicator
DOD—domestic object damage

E

ECM—electronic control module
ECSS—engineering control and surveillance system
EG—electric governor
EOCC—engineering operational casualty control
EOP—engineering operational procedures
EOSS—engineering operational sequencing system
EPCC—electric plant control console
EPCE—electric plant control enclosure

F

FO—fuel oil
FOD—foreign object damage
FSCC—fuel system control console
FSEE—free standing electronics enclosure
ft—foot

G

GB—generator (circuit) breaker
GCU—generator control unit
GG—gas generator
gpm—gallons per minute
GT—gas turbine
GTE—gas turbine engine
GTG—gas turbine generator
GTGS—gas turbine generator set
GTM—gas turbine module

H

hp—horsepower
HP—high pressure
Hz—hertz

I

IGN—ignition
IGV—inlet guide vane
in.—inches
in.H20—inches of water

K

kw—kilowatt

L

lb.—pounds
lb/min—pounds per minute
lb/sec—pounds per second
lb/hr—pounds per hour

LFV—liquid fuel valve
LO—lubricating (lube) oil
LOP—local operating panel
LOSCA—lube oil storage and conditioning assembly
LP—low pressure

M

MER—main engine room
MFC—main fuel control
mil—mils = (0.001 in.)
MIL STD—military standard
MRG—main reduction gear

N

Ni—speed voltage
N2/NPf—power turbine speed

NAVSEA—Naval Sea Systems Command
Ngg—gas generator speed (engine speed)
NSTM—Naval Ships' Technical Manual

P

PACC—propulsion and auxiliary control console
PACEE—propulsion auxiliary control electronics enclosure
PAMCE—propulsion and auxiliary machinery control equipment
PAMISE—propulsion auxiliary machinery information system equipment
PCB—printed circuit board
PCC—propulsion control console
PCS—propulsion control system
PLA—power level angle
PLCC—propulsion local control console
PLCEE—propulsion local control electronics enclosure
PLOE—propulsion local control equipment
PMA—permanent magnet alternator
PMS—planned maintenance system
POT—potentiometer
press—pressure
psid—pound per square inch differential
psig—pound per square inch gauge
PT—power turbine
Pt2—compressor inlet total pressure
Pt5.4—power turbine total inlet pressure
PTO—power takeoff

R

rpm—rotations per minute
RTD—resistance temperature detector

S

SS—ship's service
SSAS—ship's service air system
SSDG—ship's service diesel generator
SSGTGS—ship's service gas turbine generator
sup—supply
sys—system

T

T2—compressor inlet temp
T5.4—power turbine inlet temperature (LM2500)
TACH—tachometer
temp—temperature
TIT—turbine inlet temperature (GTG)
TOPS—turbine overtemp protection system

V

VSV—variable stator vane

X

XFMR—transformer
XMIT—transmit

INDEX

V

W

Printed in the United States
By Bookmasters